JN425823

A Place Where Ships are Built

Guidelines for New Shipyard Development and Design

A Place Where Ships are Built

Guidelines for New Shipyard Development and Design

author Seungwoo Kang

PNUPRESS

A Place Where Ships are Built

Guidelines for New Shipyard Development and Design

by Pusan National University Press
Copyright © Seungwoo Kang, 2024

No part of this publication may be reproduced, distributed, or transmitted in any form or by any means, including photocopying, recording, or other electronic or mechanical methods, without the prior written permission of the publisher.
However, brief quotations embodied in critical reviews and certain other noncommercial uses permitted by copyright law are allowed.

For permission requests, please write to the publisher at the address provided below.

All rights reserved.

Author Seungwoo Kang
Address 53224, 19 Jinmok-ro, Geoje-si, Gyeongsangnam-do,Republic of Korea
Email ksw076571@gmail.com

Designer Mi Ran Hwang
Publisher Pusan National University Press
Registration of publication 1983-000001 November 10, 1983.
Address 2, Busandaehak-ro 63beon-gil, Geumgeong-gu, Busan, 46241, Korea
TEL +82-51-510-1932
Fax +82-51-512-7812
Email pnupress@pusan.ac.kr

ISBN 978-89-7316-788-3(93530)

A Place
Where Ships are Built

Acknowledgements

The author wrote this book with the hope that it would serve as a reference for countries or companies seeking to create significant-scale employment opportunities and develop their national key industries through the expansion of the shipbuilding industry. Furthermore, I would like to state that while writing this book, I partially referred to the contents of overseas shipbuilding cooperation projects that I worked on with my teammates at Daewoo Shipbuilding & Marine Engineering in 2010.

In addition, the author informs readers that prior to the publication of this book, approval was obtained from Hanwha Ocean officials, who acquired DSME in 2022, in November 2023, regarding copyright and security.

The review of 'A Place Where Ships are Built'

The content of this book encompasses over 30 years of practical experience in the core areas of a world-class shipyard, as well as the accumulated know-how from consulting on new shipyard development and productivity innovation projects undertaken as part of shipbuilding collaborations with resource-rich countries worldwide.

While the shipbuilding industry is a comprehensive assembly industry that involves a wide range of fields, there is a scarcity of technological books that provide an understanding of the overall tasks handled in shipyards due to the vastness of the industry. However, this book delves into developing new shipyards based on experiential knowledge and expertise rather than relying solely on academic logic. As such, it can be a valuable resource for future generations seeking a comprehensive understanding of the shipbuilding industry. I sincerely hope this valuable book will become a blueprint for the future shipbuilding industry.

Professor Sung-chul Shin
Department of Naval Architecture and Ocean Engineering,
Pusan National University

Table of Contents

Part I.

Site Selection for a New Shipyard

Part II.

Designing of a New Shipyard

Part III.

Valuation for a New Shipyard Operation

Prologue

The author of this book has worked for more than 30 years at Daewoo Shipbuilding & Marine Engineering Co. Ltd., one of the Big 3 shipyards in Korea, and majored in naval architecture during the growth period of the Korean shipbuilding industry in the 1980s.

The purpose of this book is to share the experience of shipyard development, which is based on the expansion of DSME's production capacity and my extensive experience in shipbuilding cooperation projects with various countries abroad. It is hoped that this book will be helpful to overseas countries that plan to develop the shipbuilding industry in the future and promote the creation of significant employment opportunities for their citizens. The book covers the following three areas and is based on empirical knowledge and practical expertise, rather than academic theory.

Firstly, during the shipyard development stage, the primary aspect to consider is the selection of a suitable site with optimal location conditions for the shipyard. This is a crucial factor that directly impacts the overall cost of shipyard development. Part I will address the factors that should be considered when selecting candidate shipyard sites.

Secondly, in Part II, the book discusses the process of designing and developing a modern shipyard. It focuses on aligning the shipyard's mid- and long-term production goals and determining the appropriate production facilities for the shipyard. This process was based on the mass production of large merchant ships, with the aim of entering the international market for new shipbuilding.

Thirdly, the shipbuilding industry is a large-scale facilities industry, and a significant amount of initial development funds is required for shipyard construction. Even though the growth of the shipbuilding industry generates significant employment opportunities and has positive ripple effects on the front and rear industries, it is challenging for a single company to recoup its investment in development funds solely through shipyard operation profits. Therefore, it is crucial to review the corporate valuation based on the mid- and long-term production goals of a newly established shipyard. Furthermore, it is also important to review the additional requirements for various government support policies necessary for early normalization. This aspect will be addressed in Part III.

The content described in this book is the result of the collaborative efforts of colleagues who have worked with the author over the past 30 years, and I would like to express my gratitude to all of them in advance of publication.

A Place
Where
Ships are Built

Part I.

Site Selection for a New Shipyard

This is an example of increasing sales by extending the length of the existing dry dock to build larger ships. This can be done without interrupting production.

Part I.

Site Selection for a New Shipyard

1 A Basic Understanding of the Characteristics of the Shipbuilding Industry

The original purpose of developing the shipbuilding industry is to benefit countries that aim to promote economic development and job creation by fostering the growth of their own underdeveloped shipbuilding industry. However, it is also true that there are many cases where the basic understanding of the shipbuilding industry is inadequate.

An irrational approach, where political purposes are excessively involved or where a candidate site is determined in advance based on personal interests, has fatal flaws from the initial stage of site selection. In the end, this approach can lead to a lack of success. Therefore, decision makers must have a basic understanding of the shipbuilding industry in the initial stages of entering this sector, as it is a crucial factor.

Transition of the Global Shipbuilding Industry

Following the Industrial Revolution in England, the shipbuilding industry rapidly developed with the advent of iron ships equipped with steam engines. In the 1930s, the United States developed the arc welding technique for steel ships, which brought about a significant transformation in the shipbuilding industry. During World War II, the US Navy built large-scale Liberty Series transport ships, representing a significant milestone in shipbuilding techniques and establishing a modern approach to ship construction.

After World War II, Japanese shipyards held a prominent position in the global shipbuilding industry, particularly in the merchant ship sector, for a considerable period of time. In the 1970s, large-scale shipyards were constructed as part of the Korean Economic Development Plan's heavy industry promotion policy. The Korean shipbuilding industry played a crucial role in the Korean economy, leading in the global shipbuilding industry until the early 2010s.

Subsequently, large-scale new shipyards emerged in China, leading to a rapid decline in the shipbuilding market. This was primarily caused by a surplus of global supply capacity and indiscriminate offensives that relied on low wages and Chinese government subsidies, including financing support for shipowners. As China implemented a national policy requiring ships for domestic demand to be built in their own country, its global market share expanded rapidly. Today, Korea and China are competing fiercely for the top position in the global shipbuilding industry.

On the other hand, resource-rich countries such as Russia, Saudi Arabia, and Brazil are promoting the development of their own shipbuilding industry under the guise of the so-called Local Contents Requirement. However, it is also true that important factors such as the suitability of shipyard site conditions, productivity competitiveness, and economic feasibility (i.e., cost-effectiveness of investment) are largely overlooked.

Since the shipbuilding industry is labor-intensive and has limitations in automation, it is inevitable that the center of the global shipbuilding industry will shift to a new maritime country with cheap labor in the near future, following the footsteps of Korea and China. This industrial trend will gradually transition from low value-added ships that are relatively easy to build.

Features in terms of Shipyard Construction

The construction of new shipyards has the following basic characteristics, which result in high entry barriers for new participants in the shipbuilding market.

- A huge equipment industry involves large-scale facility investment.
- A capital-intensive industry that requires enormous financial resources.
- Vast industrial infrastructure requires a large land area, convenient transportation roads, and high-voltage power.
- Relatively deep water depth and favorable soil and seabed conditions to ensure high bearing capacity are required.

Ripple Effect on Related Industries in the Upstream and Downstream of the Shipbuilding Industry

The shipbuilding industry has a significant overall ripple effect on both upstream and downstream industries, rather than solely impacting individual companies economically.

- Downstream industries include marine energy development, shipping, fisheries, tourism, defense, etc.
- Upstream industries include the metal and steel industry, machinery industry, electricity industry, telecommunications industry, electronics industry, accommodation equipment industry, chemical industry, armaments manufacturing industry, finance and insurance industry, and more.

Features in terms of Shipyard Operations

As essential features of shipyard operations, the following points differentiate it from other manufacturing industries.

- There are limitations to standardization and automation in shipbuilding due to the nature of the industry. Unlike standardized off-the-shelf products like airplanes and automobiles, shipbuilding involves the production of customized, made-to-order products in small quantities. It is a comprehensive assembly industry that deals with super-large steel structures. Therefore, it is important to systematically establish spatial and functional integration capabilities as the core competency of a shipyard.
- Except for cases enacted by law in some countries to protect their own industries (such as the Jones Act in the US), the global shipbuilding market operates as a single market. This means that the shipbuilding market is an infinitely competitive market.
- As a labor-intensive industry, productivity is a key competitive factor for individual companies. It is represented by three competencies:price competitiveness, quality competitiveness, and delivery competitiveness. According to empirical statistics from a large global shipyard, productivity is found to be sensitive to labor skills (72%), shipyard operation systems (15%), and shipyard facilities (13%). Therefore, above all, skillful workforces are the key to competitiveness.

- The shipbuilding market and newbuilding orders worldwide exhibit significant volatility due to external factors. These factors include trends in trade and transport volume, which are influenced by the global macroeconomy. Additionally, various factors contribute to this volatility, including trends in marine resource development, such as oil and natural gas prices and global energy demand. Furthermore, shipping market conditions and shipowners' demand for fleet replacement also play a role. Recently, the implementation of new international regulations, such as environmental regulations, has a significant impact on the shipbuilding market.

- As the contract amount is relatively large compared to other industries, export ships have a high foreign exchange earning rate, such as dollars and euros.

Potential Risks in Shipyard Operations

Potential risks that threaten the sustainable management of new shipyards include the following:

- It typically takes 2 to 3 years from the award of a shipbuilding contract to its delivery. During this time, shipyards are exposed to fluctuations in exchange rates of contract currencies, such as the dollar and/or euro. Therefore, there is always the possibility that the ability to hedge currency can have a significant impact on profitability.

- Depending on the economic fluctuations in the shipbuilding industry, there are distinct periods of boom and stagnation. If there is a lack of domestic demand or planned shipbuilding orders, the production workload can experience significant volatility. Therefore, shipyards often have no choice but to accept low-priced orders during the downturn in the shipbuilding market because of the high fixed costs associated with shipyards.

- When shipowners dominate the shipbuilding market, it becomes necessary to effectively manage risks arising from contract terms and conditions, as well as unfavorable heavy tail payment conditions that pose challenges to shipbuilders.

- Sometimes, moral hazard occurs when shipyard management becomes overly focused on securing new orders instead of prioritizing profitability.
  This can happen because there is a significant time gap between the award of shipbuilding contracts and the final settlement of project profitability after delivery.

2 Understanding the Shipyard Development Processes

Even if design and construction are carried out simultaneously using a Fast Track approach, the process of building a new shipyard with a certain size or larger takes a minimum of 5 years. This includes the review stage for decision-making in entering the shipbuilding industry and completing the shipyard construction as an initial stage investment.

First, in order to obtain the final decision on whether to construct a new shipyard, it is necessary to conduct a feasibility study by a professional accounting firm and an on-site trade-off study by shipbuilding experts. In this process, the initial plan must be defined as comprehensively as possible, including the basic blueprint for shipyard construction.

After a decision is made on shipyard construction, civil and architectural engineers conduct basic data research, including collecting raw data through site investigation. Topographic surveys, geological surveys, and ocean surveys are essential for conducting design work on the selected site. In addition to this, shipbuilding experts prepare Technological Solutions for determining the size of the shipyard to be constructed or the workshops and facilities to be equipped. Based on these Technological Solutions, the basic design and detailed design for civil and architectural engineering will proceed one after another.

Based on the design results, assessments are conducted for environmental impact, traffic impact, cultural property investigation, review of disaster impact, and reporting of damage impact, among other factors. Various governmental permits, such as the designation of industrial complexes, are obtained from government and related authorities. Public hearings should also be held to obtain consent from local residents. Finally, it can be said that the detailed design for commencing the construction of the shipyard has been finalized only when the approval for the industrial complex designation is announced.

Of course, there may be significant variations in licensing requirements among different countries. Additionally, if the design is done by a foreign engineering company, conversion design must be carried out to comply with domestic standards in order to prepare approval documents. Therefore, the involvement of local engineering companies in the design process is crucial.

In order to expedite the construction of shipyards, a Fast Track approach is employed, which entails conducting detailed design and carrying out preliminary work simultaneously. These preceding works include calculating the construction quantity for the pre-qualification process of construction companies and procuring major shipyard facilities with long lead times, such as Goliath cranes. The construction period of a shipyard can vary depending on factors such as the size of the shipyard, climate conditions at the site, and the capabilities of local contractors. However, if a Korean construction company is the main contractor responsible for constructing a shipyard with the capacity to assemble 200,000 tons of steel per year, the first ship may be delivered in approximately 3 years, considering the practices of the Korean construction industry.

For reference, Figures 1 and 2 depict examples of the overall work procedures and construction period for shipyard construction in Korea.

Process & Flow Chart for Shipyard Construction *(Sample of Korean Practice)*

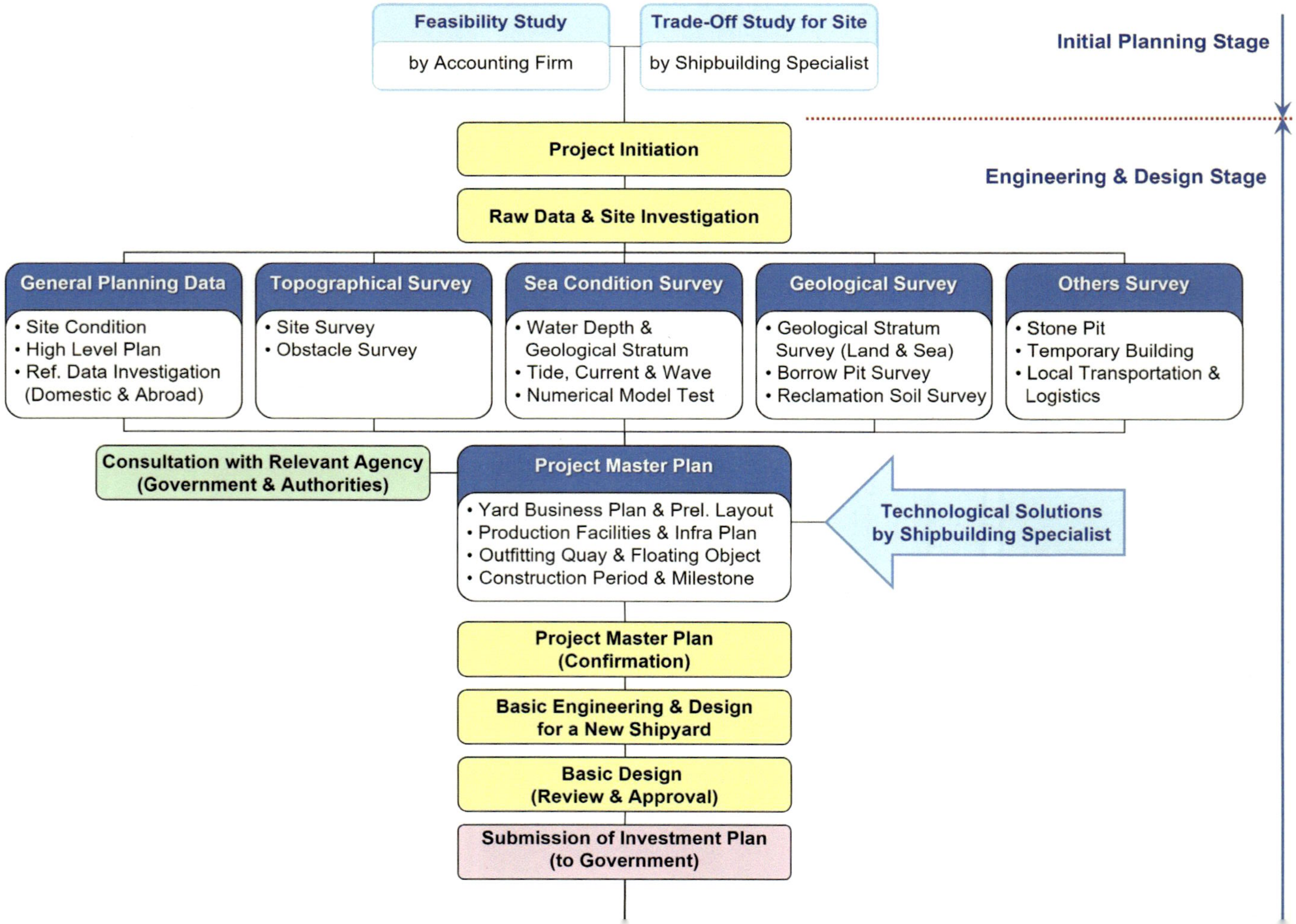

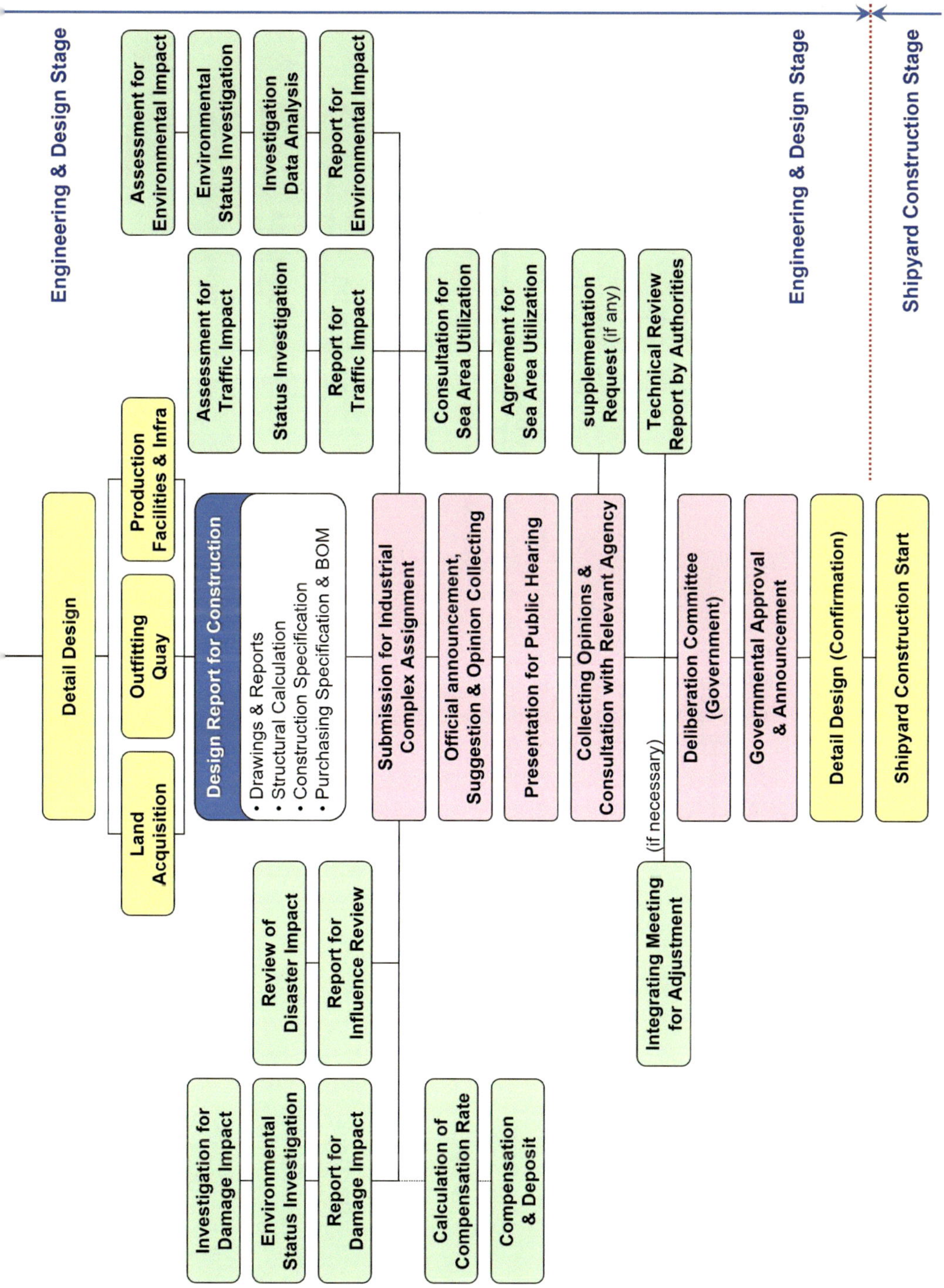

Figure 1 Process and Flow Chart for Shipyard Construction

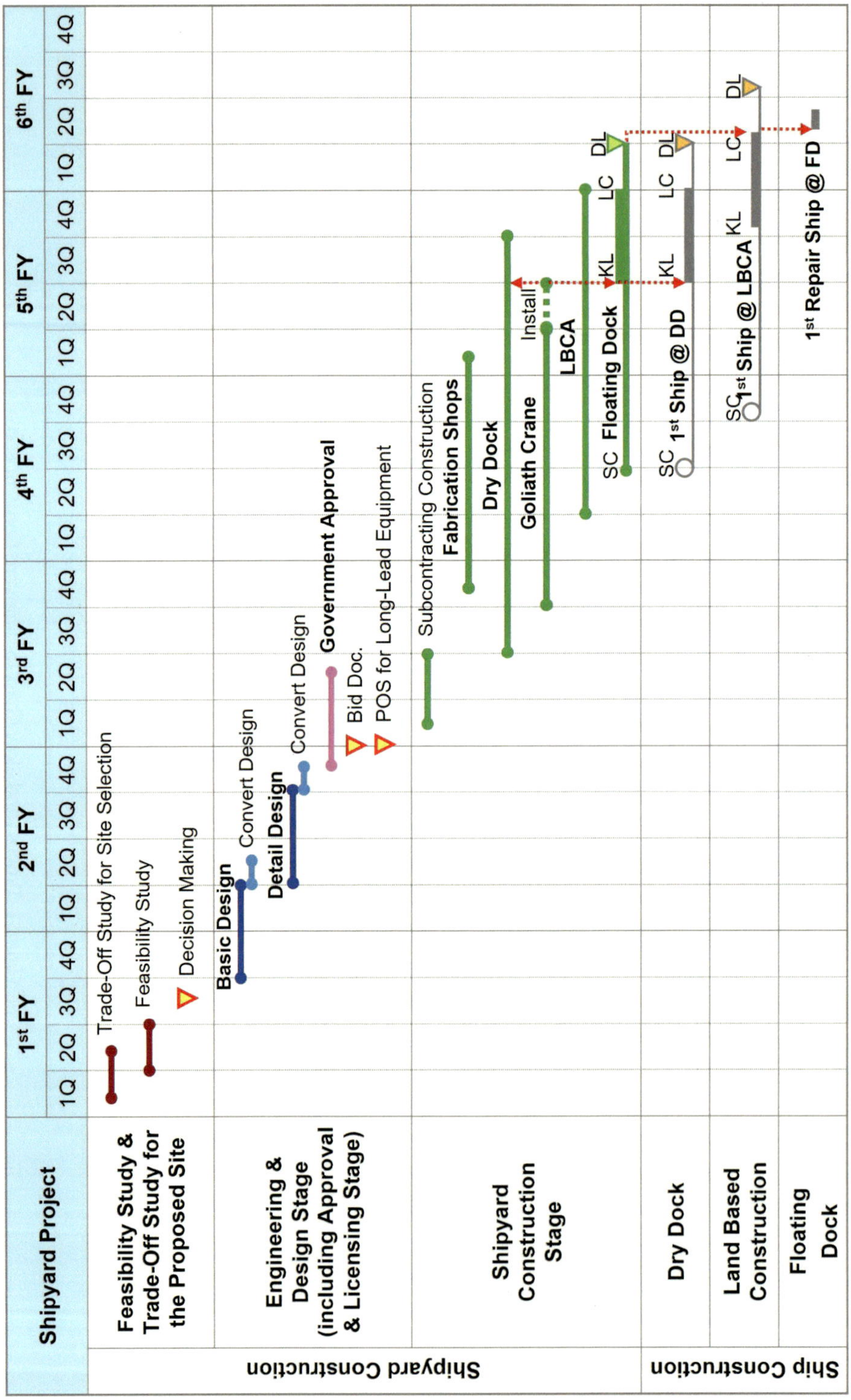

Figure 2 Typical Project Master Schedule for Shipyard Construction

2.1 Initial Planning Stage for Shipyard Construction

The primary goal during the initial planning stage of shipyard construction is to make rational decisions regarding new entry into the shipbuilding industry or expanding shipyard facilities.

To achieve this, as mentioned above, the following two tasks are usually performed:

- Feasibility Study
- Trade-Off Study for Site

2.1.1 Project Team Organization

As depicted in Figure 1, which illustrates the comprehensive process of developing a new shipyard, the success or failure of the project relies heavily on the project management capability to effectively lead the entire endeavor. This is because experts from various fields, numerous stakeholders, and complex procedures are intertwined with each other.

Of course, there may be significant differences in project management methodology depending on the timing before and after the decision-making process. However, in the preliminary stage, it is important to recognize that a rational approach and effective coordination between the business evaluation expert and the shipbuilding expert are crucial to maintain a balanced and objective perspective in the feasibility study. Although many cases of business feasibility studies for new shipyard construction are performed by world-recognized accounting firms, it is true that there are also many instances where the feasibility study is evaluated from a subjective perspective, focusing primarily on the excessively positive aspects of business feasibility. As a result, achieving positive outcomes in the future will be relatively rare. In addition, during the process of conducting a trade-off study for a potential site, adopting an irrational purpose-oriented approach, such as relying solely on a simple methodology (i.e., believing that there is always a technical solution), instead of a value-oriented approach like considering economic feasibility, can lead to unforeseen and serious issues. In these cases, unforeseen issues such as the inability to recoup the initial investment expenditure (CAPEX) or ground subsidence may occur in the future. Therefore, careful attention is needed in the composition of the team of experts who are related to the project.

In the follow-up stage after decision-making, the project management team must first familiarize themselves with the contents of "ISO 21500:2012 Guidance on Project Management," which was announced by the International Organization for Standardization (ISO) in 2012. It would be desirable to establish a collaborative project management team consisting of shipbuilding experts who will develop the blueprint for the future shipyard, incorporating technological solutions, and engineering experts who will oversee civil and architectural engineering, including government licensing requirements, based on the engineering deliverables. After that, when construction begins, a team of construction experts, including the supervision group, should also be involved in the project management team.

2.1.2 Feasibility Study

There is no need to debate whether the decision maker's decision comes first or if the business feasibility study comes first. Unfortunately, in most cases, the decision is often made without consulting an accounting firm for a business feasibility study to support it. Furthermore, it seems that in most cases, this requirement is necessary for official processes, such as obtaining board approval. Since provisional internal decision-making has already been established, and the results of the project feasibility study must be derived within a limited time, the consulting service process often adopts somewhat irrational parameters. This phenomenon generally occurs regardless of whether the project is domestic or overseas. However, in order for the top decision-maker to make the right choice, a feasibility study should be conducted using objective data and figures. It should be recognized that the decision-maker, not the staff participating in the feasibility study, bears sole responsibility for the decision to start a new business.

Although the contents of the feasibility study will vary depending on the work style of globally recognized accounting firms, they generally cover the following topics. As these parts are another area of specialization, this book does not address them in detail.

Project Overview

- Background
- Project Summary (including Production Capacity Goal)
- Shareholders

Market Analysis

- Macro Environment (including pricing for each product)
- Demand Forecast for Vessels and Offshore Products
- Competitor Analysis

Business Plan

- Project Schedule (including Shipbuilding Schedule and Overall Timeframe)
- JV (Joint Venture) Development Plan
- Role and Responsibility
- JV Incorporation
- Organizational Structure
- Yard Construction
- Human Resources and General Affairs (including Training Program)
- Marketing
- Engineering
- Procurement
- Production
- Management (including QA/QC system)

Financial Analysis

- Business Model Overview
- Projection Flow
- Assumptions (including Productivity Improvement Plan)
- Feasibility Study Results (including CAPEX Estimates)
- Forecasted Financial Statements
- Scenario Analysis (i.e., Cash Flow by Scenario)
- Risk Analysis

Strategy and Vision

- SWOT Analysis
- Business Strategy
- Vision and Goal Roadmap
- Benefits and Contributions to the Country

2.1.3 Trade-Off Study for Candidate Sites

It is necessary to roughly determine the size of the new shipyard in question through market research before conducting a trade-off study for potential candidate sites for the shipyard.

Business Type and Scale of the New Shipyard (i.e., Goals of the Initial Stage)

Based on the minimal and brief market research presented in Table 1, it is possible to make an estimation regarding the type and scale of the business to be established. The following factors can be considered:

- Type of new shipyard business: New shipbuilding, ship repair, or both new shipbuilding and repair.
- Product lines: Cargo ships, passenger ships, offshore platforms, naval and coast guard vessels, and more.
- Production capacity: Number of new ships delivered per year (or the number of repairs) or the tonnage of steel cut per year.
- The ratio of exports to domestic demand, etc.

<table>
<tr><th rowspan="2">Type of Product</th><th rowspan="2">Capacity or Size</th><th colspan="5">Number of Project Forecasts</th></tr>
<tr><th>Total</th><th>1st FY</th><th>2nd FY</th><th>~</th><th>10th FY</th></tr>
<tr><td colspan="7">Merchant Ship</td></tr>
<tr><td>Bulk Carrier</td><td></td><td></td><td></td><td></td><td></td><td></td></tr>
<tr><td>Crude Oil Tanker</td><td></td><td></td><td></td><td></td><td></td><td></td></tr>
<tr><td>Container</td><td></td><td></td><td></td><td></td><td></td><td></td></tr>
<tr><td>Liquefied Natural Gas Carrier</td><td></td><td></td><td></td><td></td><td></td><td></td></tr>
<tr><td>Liquefied Petroleum Gas Carrier</td><td></td><td></td><td></td><td></td><td></td><td></td></tr>
<tr><td>Pure Car & Truck Carrier</td><td></td><td></td><td></td><td></td><td></td><td></td></tr>
<tr><td colspan="7">Offshore Product</td></tr>
<tr><td>Float. Product Storage & Offloading</td><td></td><td></td><td></td><td></td><td></td><td></td></tr>
<tr><td>Float. Storage Regasification Unit</td><td></td><td></td><td></td><td></td><td></td><td></td></tr>
<tr><td>Semi-Rig / Jack-Up Rig</td><td></td><td></td><td></td><td></td><td></td><td></td></tr>
<tr><td>Drillship</td><td></td><td></td><td></td><td></td><td></td><td></td></tr>
<tr><td>Floating Production Unit</td><td></td><td></td><td></td><td></td><td></td><td></td></tr>
<tr><td>Fixed Platform</td><td></td><td></td><td></td><td></td><td></td><td></td></tr>
<tr><td colspan="7">Passenger Ship</td></tr>
<tr><td>Ro-Ro Passenger Ferry</td><td></td><td></td><td></td><td></td><td></td><td></td></tr>
<tr><td>Passenger Ferry</td><td></td><td></td><td></td><td></td><td></td><td></td></tr>
<tr><td>Cruiser</td><td></td><td></td><td></td><td></td><td></td><td></td></tr>
<tr><td colspan="7">Special Ship</td></tr>
<tr><td>Offshore Supply Vessel</td><td></td><td></td><td></td><td></td><td></td><td></td></tr>
<tr><td>Fishing Vessel</td><td></td><td></td><td></td><td></td><td></td><td></td></tr>
<tr><td>Tugboat</td><td></td><td></td><td></td><td></td><td></td><td></td></tr>
<tr><td colspan="7">Naval Ship</td></tr>
<tr><td>Destroyer</td><td></td><td></td><td></td><td></td><td></td><td></td></tr>
<tr><td>Frigate</td><td></td><td></td><td></td><td></td><td></td><td></td></tr>
<tr><td>Offshore Patrol Vessel</td><td></td><td></td><td></td><td></td><td></td><td></td></tr>
<tr><td>Submarine</td><td></td><td></td><td></td><td></td><td></td><td></td></tr>
</table>

Table 1 Potential Projects for New Shipbuilding and/or Ship Repair

Based on the initial stage business strategy mentioned above, shipbuilding experts and civil and architectural engineering experts collaborate to compare and review the following factors for potential new shipyard sites. They then select a site that offers a comparative advantage.

Comparison of the Current Status of Shipbuilding Industry Infrastructure

- Electricity Supply
- Water Supply
- Gas and Fuel Oil Supply
- Shipbuilding Equipment and Materials Supply Chain
- Manpower Supply and Productivity, etc.

Comparison of Social Infrastructure Status

- Communication Network
- Transportation
- Residential Complexes
- Environmental Regulations
- Administration Services, etc.

Comparison of Shipbuilding Industry Support Policies

- Site Acquisition
- Taxation System
- Industrial Infrastructure Support
- Government Approval and Clearance, etc.

Comparison of Sea Environmental Conditions

- Water Depth and Nautical Charts

- Tidal Range and Current Speed
- Wave Height
- Seabed Geological Survey Data, etc.

Comparison of Land Environmental Conditions

- Site Circumstance
- Ground Leveling and Reclamation
- Geological Survey and Nature of Soil
- Earthquake
- River, etc.

Comparison of Climate and Environmental Conditions

- Temperature, Rainfall, Snowfall, and Humidity
- Wind, Tropical Storms, and Tsunamis, etc.

Key considerations and details of the comparison of these new shipyard candidates are covered in Chapter 3.

2.2 Basic and Detailed Design Stages for Shipyard Construction

After deciding to enter the shipbuilding industry market, the next step is to proceed with the basic design and detailed design for construction licensing. To do so, shipbuilding experts must first develop Technological Solutions. Based on this solution, civil and architectural engineering companies are then tasked with performing basic and detailed design.

2.2.1 Role of Shipbuilding Specialists

A technological solution, prepared in advance by shipbuilding experts, incorporates modern shipbuilding methods that align with the business portfolio and production goals of the new shipyard. Based on this solution, civil and architectural engineering companies are requested to submit basic and detailed designs for construction licensing. However, it is an unfortunate reality that logically developed references or samples of these technological solutions do not exist. This is the reason why this book was created.

The technological solution, established for the first time in the world during the OOO project in 2010, covers the following seven topics:

- General Shipbuilding Process
- Goals of a New Shipyard
- Capacity of Each Production Process
- Layout of a New Shipyard
- Shipyard Equipment and Facilities
- Training and Technology Transfer
- IT Infrastructure

Although the design guidelines for each subject are not based on academic logic, they can serve as valuable references when constructing a new shipyard. These guidelines are written based on empirical shipyard operational knowledge from a globally recognized top-tier shipyard. Additionally, someone may establish more comprehensive shipyard design guidelines based on this in the future.

The details of the contents to be considered for each subject will be discussed in "Part II: Designing a New Shipyard".

2.2.2 Role of Civil and Architectural Specialists

Since the civil and architectural engineering fields differ from the shipbuilding industry, they will not be discussed here. However, it is important to note that both basic and detailed design processes must be completed in order to obtain permits from the country where the shipyard is being constructed. It is also crucial for local engineering companies to participate in this process. Otherwise, after the design is completed, local engineering companies must conduct a conversion design to comply with the relevant local regulations. Therefore, this period should be taken into account as an additional factor when establishing the project engineering schedule.

In the case of constructing a shipyard using the Fast Track method, where design and construction work are carried out simultaneously, shipbuilding experts place timely purchase orders for equipment with long lead times, such as Goliath cranes, and partially operate the shipyard during the shipyard construction process.

In addition, if it is not an EPC turn-key contract method that includes all aspects of Engineering, Procurement, and Construction, the basic and detailed design companies generally need to prepare bidding notice documents and pre-qualification documents in advance for selecting construction contractors.

2.3 Shipyard Construction Stage

As shipyard construction takes a significant amount of time from design to completion, it is crucial to establish a construction schedule that allows for the partial operation of the shipyard during construction. Since many steel structures need to be manufactured during the construction process, it is more efficient to establish steel stock yards, cutting workshops, assembly workshops, and blasting and painting workshops, even if only partially. Additionally, during the shipyard construction process, it is necessary to consider prioritizing the provision of outdoor workspaces, such as pre-erection areas, to enable the assembly and erection of large structures like Goliath cranes and dock gates. Furthermore, it is desirable to establish a construction schedule that allows for the launch of the first ship once the dry dock is completed.

Another critical aspect of the shipyard construction phase is the construction supervision team. It is common for a qualified construction supervision team to perform its roles at basic and detailed design companies that are familiar with the project's design details. The construction supervision team must manage progress, budget, quality, administration, and completion work with the objective of achieving optimal quality, timely completion, and staying within budget.

3 Detail Considerations for Proposed Site Selection

3.1 Infrastructure of the Shipbuilding Industry

To evaluate the condition of the local shipbuilding industry infrastructure, Table 2 presents items for a comprehensive comparison and review of various factors related to the local shipbuilding industry infrastructure.

Electricity Supply

A shipyard is an industry that typically consumes a significant amount of electricity. If the production capacity of a shipyard is larger than that of a medium-sized shipyard, it is desirable to directly connect the transmission power line from the transmission tower to the main substation of the shipyard. Additionally, the stability of voltage and frequency fluctuations must be properly ensured. If a duplicate main transmission power line is installed to the shipyard substation, it can prevent blackouts caused by external forces such as typhoons. Korean shipyards located in typhoon-prone areas have installed double main transmission power lines as a precautionary measure following a major transmission tower collapse that resulted in significant damage.

Water Supply

In addition to providing drinking water for the residents of the shipyard and the crew on board the sea trial vessel, fresh water and/or graywater is also needed for hydraulic testing of the hull structure carried out in the dry dock. Therefore, if there are no rivers or streams with abundant water sources near the shipyard, or if there is no reservoir for storing rainwater in the shipyard, it may be necessary to develop an underground water source.

Gas and Fuel Oil Supply

Carbon dioxide is used for CO_2 welding to minimize thermal deformation that affects the hull materials, while argon gas is used for welding non-ferrous pipes. Oxygen and acetylene gases are used for oxyfuel gas cutting, while a diverse range of gases including LNG, LPG, helium, and nitrogen must be supplied properly. It is also necessary to supply duty-free oil of the appropriate type for sea trial vessels.

Shipbuilding Equipment and Materials Supply Chain

The steel industry, which can provide shipyards with steel plates in a stable and timely manner, is one of the most crucial factors for their competitiveness. In Korea, thicker plates for shipbuilding can be supplied in the desired size within 30 to 40 days. However, if there are numerous restrictions on the specifications of steel plates produced at steel mills or if steel plates can only be transported by land transport trailers or railroads, the availability of steel plate options may be severely limited. This limitation can have a detrimental impact on the productivity of shipyards.

As shipyards develop, the supply chain for shipbuilding equipment, including engines, generators, propellers, navigation and communication devices, and accommodation facilities, as well as raw materials like steel plates, cables, pipes, and paints, must also evolve. This factor is crucial in the development of the shipbuilding industry. In Korea, the localization rate of equipment for general merchant ships exceeds 90%.

Manpower Supply and Productivity

Since the shipbuilding industry is labor-intensive, the availability of engineers and labor is crucial for maintaining price competitiveness. Comparison of working conditions in a trade-off study of potential shipyard sites within a country may not hold much significance. However, for future shipyard valuations, wages, working conditions, and annual available working hours are crucial basic information.

In addition, the most crucial factor is the availability of data that can objectively compare labor productivity. Even if it is not a ship of the same class being built, reference data on the index of production working hours per steel net ton (i.e., man-hour/steel net ton) of the local shipyard and the construction period from steel cutting to delivery by ship type are very important pieces of information for assessing local productivity.

This is one of the most significant factors in developing a mid- and long-term plan for manpower supply and demand, which is based on labor productivity.

On the other hand, it is necessary to keep in mind that a candidate site that is excessively far from a densely populated residential area may require the development of a residential complex in addition to the construction of a shipyard.

Scope of Preliminary Site Investigation	Status
1. Infrastructure of the Shipbuilding Industry	
1.1 Electricity Supply	
1.1.1 Electric Power Supply Capacity - i.e., 154 kV, 50 or 60 Hz, 120,000 kW	
1.1.2 Allowance for Electric Power Supply	
1.1.3 Distance from Primary Substation	
1.1.4 Unit Price per kW·h	
1.1.5 Fluctuation Range of Voltage and Frequency	
1.1.6 Number of Electric Power Supply Lines - i.e., Single or Duplicate	
1.1.7 Frequency of Electric Power Supply Blackout	
1.2 Water Supply	
1.2.1 Capacity of Water Supply per Day	
1.2.2 Distance from a Storage Reservoir	
1.2.3 Quality of Water Service	
1.2.4 Underground Water Resources	
1.2.5 Unit Price per ton	
1.3 Gas and Fuel Oil Supply	
1.3.1 LPG Supply Capacity per Year and Unit Price	
1.3.2 CO_2 Gas Supply Capacity per Year and Unit Price	
1.3.3 Argon Gas Supply Capacity per Year and Unit Price	

Scope of Preliminary Site Investigation	Status
1.3.4 Oxygen Gas Supply Capacity per Year and Unit Price	
1.3.5 Ethylene Gas Supply Capacity per Year and Unit Price	
1.3.6 Helium Gas Supply Capacity per Year and Unit Price	
1.3.7 Nitrogen Gas Supply Capacity per Year and Unit Price	
1.3.8 Gasoline Supply Capacity per Year and Unit Price	
1.3.9 Paraffin Oil Supply Capacity per Year and Unit Price	
1.3.10 Bunker-A Oil Supply Capacity per Year and Unit Price	
1.3.11 Bunker-C Oil Supply Capacity per Year and Unit Price	
1.4 Shipbuilding Equipment and Materials Supply Chain	
1.4.1 Supply Chain of Domestic Steel Mill (Plate & Section Profile) - Annual Capacity - Manufacturing Standard Sizes - Logistics (Land and Marine Transportation)	
1.4.2 Domestic Vendor List for Marine Equipment and Materials (Registered with a Classification Society	
1.5 Manpower Supply and Productivity	
1.5.1 Rights and Role of Labor Unions	
1.5.2 Types of Hiring Contracts	
1.5.3 Flexicurity in the Labor Market	
1.5.4 Labor Supply	
1.5.5 Availability of Outsourced Manpower	
1.5.6 Status of - Technical School - Engineering College including Marine Engineering	
1.5.7 Wage Levels in the Shipbuilding Industry - Average Wage Level of Labor - Average Wage Level of Engineers - Average Wage Level of Office Workers	
1.5.8 Annual Wage Increase Rate over a Decade	

Scope of Preliminary Site Investigation	Status
1.5.9 Insurance, Supplementary Living Allowance, etc. (Additional Personnel Expenses)	
1.5.10 Working Hours - Working Hours in a Day - Average Overtime in a Day - Annual Working Days	
1.5.11 Productivity of Shipbuilding (Steel Net Ton/M-Hour) - Parallel Mid Body of a Typical Vessel, such as a Bulk Carrier - Fore and Aft Body of a Typical Vessel, such as a Bulk Carrier - Typical Topside Module	
1.5.12 Productivity of Shipbuilding (Building Period) - Typical Vessel, such as a Bulk Carrier (Contract Award ~ Steel Cutting ~ Keel Laying ~ Launching ~ Delivery) - Typical Offshore Products, such as Semi-Rigs - Typical Topside Module	

Table 2 Site Investigation Checklist for Shipbuilding Infrastructure

3.2 Social Infrastructure

Table 3 presents a comprehensive breakdown of items for comparing and reviewing social infrastructure.

Communication Network

Wired and wireless communication infrastructure has a close correlation with the automation of production facilities, particularly in new shipyards. For example, it is being developed in various forms, such as multiple types of anti-collision safety devices that rely on crane position information, warehouse automation, and remote automatic operation of a steel stock yard. Recently, the concept of Smart Shipyard (also known as Shipyard 4.0), which involves the extensive integration of rapidly advancing ICT (Information and Communication Technology) into the shipbuilding industry, is currently being implemented. To achieve this goal, it is necessary to establish a robust wired and wireless communication network that can efficiently process vast amounts of information.

In addition, frequencies must be allocated by the relevant authorities to enable not only short-distance wireless communication for exchanging messages between workers for production operations in the shipyard, but also very high-frequency (VHF) wireless communication for direct communication with the sea trial vessel.

Transportation

The logistics networks of roads, railways, ports, and airways must be established to ensure efficient transportation of shipbuilding equipment and materials from both domestic and overseas supply chains. The impact of this on delivery competitiveness should never be overlooked.

In addition, there have been instances of overseas shipyard development where customers have had to visit the shipyard for several hours and have experienced inconvenience. This inconvenience may cause customers to naturally avoid visiting again. Therefore, if necessary, helipads within or adjacent to the shipyard should also be considered.

Residential Complexes

In cases where there is no existing large-scale residential complex in the vicinity of a potential new shipyard site that can provide an adequate labor force, it is often necessary to develop a residential complex alongside the shipyard. In addition, if a shipyard specializing in ship repairs is constructed in a location far from residential areas, there may be instances where crew members avoid docking at this shipyard due to the inconvenience of finding accommodations during the two-week repair period. Therefore, complementary measures for this need to be fully considered in advance.

Environmental Regulations

Since the construction of a new shipyard results in artificial changes to the natural environment, including configuration changes to both land and sea, it is crucial to ensure compliance with all local environmental laws and regulations.

In addition, sufficient consideration must be given to the shipyard's basic and detailed design process to enable a legal response to various environmental hazards, such as noise, air pollution, sewage and wastewater, and industrial waste generated during the shipyard's operations process.

Administration Services

The Customs, Immigration, and Quarantine (CIQ) Service is required by the relevant authorities for the sea transportation of shipbuilding equipment and materials imported from overseas, as well as for the arrival and departure of ship repairs, and the departure after the delivery of new ships. They should provide immediate administrative services within the shipyard or in adjacent areas. In addition, a bonded area must be allocated within the shipyard for customs clearance of imported goods.

The environmental impact assessment typically takes the longest time when it comes to obtaining shipyard construction permits, and the evaluation process can last for one year or even longer. Since there are often provisions related to various benefits for the development of industrial complexes and the attraction of foreign direct investment, it is crucial to review the relevant regulations set by the local administrative authorities.

Scope of Preliminary Site Investigation	Status
2. Social Infrastructure	
2.1 Communication Network	
2.1.1 Status of Wire Communication Service	
2.1.2 Speed of Internet Connection and Amount of Internet Traffic	
2.1.3 Cell Phone Service	
2.1.4 Available Frequencies for Wireless Communication	
2.1.5 Distance Limit of Wireless Communication	
2.1.6 Available (Adjacent) Radio Relay Station	
2.1.7 Unit Price Rate - Wired Communication - Internet - Cell Phone - Rental Fee for Radio Waves	
2.2 Transportation	
2.2.1 Construction Plan for Highway Access to the Site	
2.2.2 Distance from the Site to the Express Highway	
2.2.3 Distance from the Site to the Domestic & International Airports	
2.2.4 Distance from the Site to the Railway Station	
2.2.5 Distance from the Site to the Trading Port	
2.2.6 Availability of Private Heliport Operations	
2.3 Residential Complexes	
2.3.1 Adjacent Cities - Distance from the Site - Population of Adjacent Cities	
2.3.2 Distance from Residential Areas to the Site	
2.3.3 Status of Medical Institutions	
2.3.4 Status of Banking Institutions	
2.3.5 Status of Convenient Life Facilities	
2.3.6 Housing Costs	

Scope of Preliminary Site Investigation	Status
2.4 Environmental Regulations	
2.4.1 Environmental Regulations - Polluted Water, Oily Water, Industrial Waste, Air Pollution, Noise	
2.5 Administrative Services	
2.5.1 CIQ Service at Site (Customs, Immigration, and Quarantine)	
2.5.2 Designation of Bonded Area at Site	
2.5.3 Assessment of Environmental Impact - Required Period	
2.5.4 Establishing a Joint Venture - Process - Required Period	

Table 3 Site Investigation Checklist for Social Infrastructure

3.3 Government Support Policy for the Shipbuilding Industry

It has been mentioned before that it is practically difficult to recover the investment unless the government of the country provides sufficient support when a local domestic company enters the shipbuilding industry or when a foreign company makes a new investment in the local area. Since the ripple effect on the upstream and downstream industries is enormous, extending beyond the business feasibility of the shipyard as an individual company, it is essential for new investors to receive maximum support from the government.

Table 4 presents a comprehensive breakdown of items for comparing and reviewing government support policies for the shipbuilding industry.

Site Acquisition

Depending on the national system, there are cases where foreign investors are only granted the right to use the land for a certain period of time, rather than ownership of the land. Therefore, the evaluation of land prices must take into account this distinction. As mentioned later, it is important to note that the cost of land per unit area should not be evaluated solely based on the site area of the land and seabed area occupied by the new shipyard. This is because it can result in significant variations in the cost of land development depending on the geological conditions.

In addition, the distance from the shipyard quayside to the international voyage route, anchorage, and official sea trial area is also a factor to consider when comparing and evaluating the safety of the navigational route and the cost of shipyard construction.

Taxation System

Since the shipbuilding industry is capital-intensive, it is not realistic to construct new shipyards without proper tax support from the central or local governments. Therefore, it is necessary to closely examine all tax standards associated with shipyard construction and operation, including social security tax, acquisition tax, property tax, value-added tax, import duty, and corporate tax. Additionally, it is important to consider the various tax benefits that can be provided by both the central and local governments.

If excessively positive profitability is predicted in the feasibility study for overseas direct investment or joint venture, there is a possibility that difficulties may be encountered in accounting audits by tax authorities during the subsequent shipyard operation process. Therefore, it is essential to seek advice and analysis on local tax laws from a tax consultant.

Industrial Infrastructure Support

The social infrastructure, including electrical power supply, roads, water supply, and communication networks, should be directly supported by the central or local governments for the new shipyard site.

Additionally, if the candidate site for a shipyard does not have terrain that can naturally shelter the inflow of waves from the open sea, the construction of a breakwater is inevitable. However, it is necessary to receive direct support from the government when a breakwater is needed because the construction of a breakwater can incur significant costs depending on the condition of the seabed.

Government Approval and Clearance

Since there are many conflicts of interest involved in the development of large-scale industrial complexes, such as new shipyards, the various permit activities generally take a long time. As this process has a significant impact on the overall construction period, it is essential to obtain full support from the relevant authorities.

Additionally, a partial use permit must be granted to allow for partial operation at each stage of the shipyard construction process. After the construction is completed, the shipyard quay should be designated as a trading port for the maritime transportation of bulky equipment, such as main engines, which cannot be transported by land.

Scope of Preliminary Site Investigation	Status
3. Government Support Policy for the Shipbuilding Industry	
3.1 Site Acquisition	
3.1.1 Total Available Area - Production Facilities - Offices, Parking Lot, etc.	
3.1.2 Site Acquisition Cost	
3.1.3 Distance from the Quay to the Sea Route	
3.1.4 Distance from the Site to the Official Sea Trial Area	
3.1.5 Distance from the Site to the Anchorage Area	
3.2 Taxation System	
3.2.1 Local Taxes, Duty & Charges	
3.2.2 Tax Breaks from Local Government	

Scope of Preliminary Site Investigation	Status
3.3 Industrial Infrastructure Support from the Local Government	
3.3.1 Electric Power Supply to the Site	
3.3.2 Highway Construction from Residential Areas and/or Cities	
3.3.3 Communication Network Service to the Site	
3.3.4 Water Service to the Site	
3.3.5 Breakwater Construction, if necessary	
3.4 Government Approval and Clearance for Shipyard Construction	
3.4.1 Types of Government Clearance and Required Periods - For Site Acquisition - For Civil Construction - For Sea Area Use	
3.4.2 Types of Government Clearance and Required Periods - Partial Operation by each Workshop during Yard Construction - Full Operating Commencement of the Shipyard	
3.4.3 Designation of Trading Port (Shipyard Quay Area)	

Table 4 Site Investigation Checklist for Government Support Policy

3.4 Sea Environmental Conditions

Table 5 presents a comprehensive breakdown of items for comparing and reviewing marine environmental conditions.

Water Depth and Nautical Charts

One of the favorable conditions for selecting a candidate site for a new shipyard is to first achieve a balance between the soil generated during the land flattening process and the amount of landfill required for sea reclamation, and also when an appropriated quay water depth is secured without separate dredging work. To accomplish this, a bathymetric survey must first be conducted from the shoreline of the potential shipyard site to the construction area.

In addition, a dense bathymetric survey should be conducted to determine the optimal location of the turning basin in front of the shipyard, taking into account the turning of both incoming and outgoing vessels. These surveys are also necessary to determine the navigational route for ships approaching existing international shipping lanes from the shipyard.

Tidal Range and Current Speed

In general, most shipyards experience a tidal range of 2 to 3 meters. However, if the tidal range exceeds this level excessively, it cannot be considered suitable for a shipyard. This is because the height from sea level to the shipyard quayside and the water depth at the quay must be adjusted to account for the tidal range. In addition, when a ship is launched into the sea using a floating dock after being constructed in a land-based construction area, the excessive tidal range can significantly restrict the design and operation of the floating dock.

On the other hand, even if a shipyard is constructed downstream of a river, it may not be considered an ideal location if there is excessive fluctuation in the flow rate depending on the season.

Waves

The potential site for a new shipyard should be protected from waves and swells, either by natural terrain or artificial breakwaters. In addition to determining the necessity of a breakwater, it is essential to collect observational data on waves and swells in the potential site area for the design of dock gates and quays.

On the other hand, in order to conduct official sea trials before delivering a ship, it is necessary to establish a designated sea trial area on the nautical chart. This area should have appropriate water depth and fishing restrictions to ensure the safe operation of the sea trial.

Seabed Geological Survey

According to various experiences in shipyard development, it is evident that while it is not technically impossible, constructing a shipyard in an area with an excessively deep seabed rock layer is not economically feasible due to the significant costs of civil construction. There have been many cases in which this aspect has been overlooked by project owners or managers, which may result in the project being cancelled in the future.

As another example, in the case of an industrial complex constructed at a relatively low cost in an area with a deep coastal sediment layer, it is also often observed that differential subsidence occurs for a significant duration even after the completion of construction. Therefore, before making a final decision on a candidate site for a new shipyard, it is desirable to conduct preliminary verification, even in the form of a sample survey, rather than a comprehensive survey.

Scope of Preliminary Site Investigation	Status
4. Sea Environmental Conditions	
4.1 Water Depth and Nautical Charts	
4.1.1 Nautical Charts	
4.1.2 Close Spaced Water Depth Chart from the Existing Shoreline	
4.2 Tidal Range and Current Speed	
4.2.1 Tidal Range - Tidal Range Table - Maximum and Average Tidal Change Annually - Approximate Lowest Low Water Level - Approximate Highest High-Water Level - Mean Sea Level	
4.2.2 Current Speed and Direction - During High Water (Tide) and Low Water	
4.3 Waves	
4.3.1 Design Criteria for Dock Gate and Quay - Significant Wave Height ($H^{1/3}$) - Wave Period and Direction	

Scope of Preliminary Site Investigation	Status
4.3.2 Swell Occurrence and Swell Data	
4.3.3 The Necessity of a Breakwater	
4.3.4 Environmental Conditions of the Official Sea Trial Area - Wind Velocity and Direction - Wave Height and Direction - Current Speed and Direction - Water Depth	
4.4 Geological Survey Data of Seabed	
4.4.1 Seabed Topography	
4.4.2 Nature of Soil beneath Seabed	
4.4.3 Depth of Rock Stratum beneath Seabed	
4.4.4 Sedimentary Stratum	
4.4.5 Report of Geological Survey beneath Seabed	
4.4.6 Estimated Amount of Dredging - Securing the Water Depth Requirement of the Quayside - Floating Dock Operation	
4.4.7 Unit Price Rate for Dredging	

Table 5 Site Investigation Checklist for Sea Environment

3.5 Land Environmental Conditions

Table 6 presents a comprehensive breakdown of items for comparing and reviewing land environmental conditions.

Site Circumstances

As an important aspect to consider in the process of reviewing how to create a site for a new shipyard, it should be noted that if the shipyard site is located far from an existing residential complex, a new residential complex should also be developed. Additionally, when considering the development of an industrial complex, the location of the shipyard should be determined by examining its connection with other industrial complexes.

Ground Leveling and Reclamation

The most effective way to secure the elevation of the shipyard land site and the required water depth at the end of the quay wall from the Approximate Lowest Low Water Level (ALLW), which serves as the baseline for water depth on the chart, is to achieve a balance between the amount of land excavation and coastal reclamation.

On the other hand, since each country exhibits significant variation in the cost of civil construction works, it is essential to verify the unit price for each type of local construction work. This is a crucial factor in accurately estimating the initial budget.

Geological Survey and Nature of Soil

As with the seabed geological survey, conducting a land geological survey and investigating strata are also crucial. The value of ground bearing forces required varies greatly depending on the shipyard's layout. In particular, the location and number of boreholes for geological and stratum investigation should be determined, considering the position of the keel block in the dry dock and the location of the Goliath crane rail where significant loads are applied.

It is recommended to conduct a minimum number of sample surveys of the land and sea areas before making the final decision on selecting a shipyard candidate site.

Earthquakes

Earthquakes can cause significant damage not only to building structures but also to shipbuilding facilities, including cranes of all kinds, trestles for supporting hull blocks, and piping systems in the shipyard. Therefore, it is necessary to check the seismic design criteria of the country, as well as statistical data on the frequency and magnitude of earthquakes in the region.

Rivers

In the case of a shipyard constructed downstream of a river, it is important to gather statistical data on changes in water flow and conduct a preliminary study on the probability of flooding. Additionally, if there is a stream or tributary near the boundary of the shipyard site, there is a possibility of ongoing sedimentation in the shipyard waters.

Scope of Preliminary Site Investigation	Status
5. Environmental Conditions of the Land	
5.1 Site Circumstances	
5.1.1 Area of - Ground Area - Reclamation Area - Greenfield Area	
5.1.2 New Town Planning at Close Range	
5.1.3 Developing a Plan for an Industrial Complex	
5.1.4 Topographical Map (1:5,000, 1:25,000, 1:50,000 Scale)	
5.2 Ground Leveling and Reclamation	
5.2.1 Height of Ground Level from Datum Line (Approximate Lowest Low Water Level)	
5.2.2 Estimation of Soil Removal Amount	
5.2.3 Estimated Reclamation Amount	
5.2.4 Capacity of Ready-Mixed Concrete Supply	
5.2.5 Unit Price Rate - For Soil Removal - For Reclamation - Reinforcing Sand, Gravel, and Rock - Civil Construction Labor Wage	

Scope of Preliminary Site Investigation	Status
5.3 Geological Survey and Nature of Soil	
5.3.1 Nature of Soil	
5.3.2 Depth of Rock Stratum	
5.3.3 Geological Survey Report	
5.4 Earthquake	
5.4.1 Statistical Data on Earthquakes	
5.4.2 Aseismatic Design Criteria	
5.5 River	
5.5.1 Probability of River Overflow	

Table 6 Site Investigation Checklist for Land Environment

3.6 Climate Conditions

Table 7 presents a comprehensive breakdown of items for comparing and reviewing climatic environmental conditions.

Temperature, Rainfall, Snowfall, and Humidity

Temperature, rainfall, snowfall, and humidity have a significant impact on the productivity of outdoor work. For instance, extreme high temperatures of 30°C or higher can lead to heatstroke among outdoor workers. Similarly, extremely low temperatures of -2°C or lower necessitate preheating for outdoor welding. Additionally, outdoor work is limited during periods of rainfall more than 5 mm. In the case of heavy snowfall, both internal and external logistics are hindered without snow removal. Additionally, high humidity can lead to immediate corrosion if the steel surface is not promptly painted after blasting. Therefore, these weather conditions directly affect productivity.

In regions with consistently hot or cold climates, it is essential to have air conditioning and heating systems in the workshop building. However, special consideration should be given to ventilation facilities, as the fumes produced during welding can have an impact on the health of employees.

Therefore, it is necessary to carefully examine the effects of these climatic environments when setting mid- and long-term productivity targets based on predictions of future productivity improvements.

Wind, Tropical Storms, and Tsunamis

Wind speed can directly affect productivity, especially when it exceeds a certain level. In general, when the wind speed reaches 10 meters per second or higher, welding defects may occur, even if proper windproof measures are taken during outdoor welding. Additionally, the use of outdoor cranes is prohibited when the wind speed exceeds 16 meters per second, and outdoor work is fully suspended when the speed exceeds 24 meters per second.

Depending on the frequency and intensity of tropical storms, such as typhoons, it is crucial to assess and establish the design criteria for the spacing of anchoring pieces for building cladding, wind load on cranes, and mooring facilities like fenders and bitts on quay walls. Some shipyards in Korea experienced significant damage from Typhoon Maemi in 2003. However, after thorough revision and strengthening of the internal design standards, there have been no further instances of similar damage.

When calculating the wind load on a ship anchored at the quay, it is advisable to use the average wind speed for 1 minute rather than the 10-minute average wind speed provided by the Meteorological Administration since the ship is a buoyant object floating on the sea.

Scope of Preliminary Site Investigation	Status
6. Weather Conditions	
6.1 Temperature, Rainfall, Humidity, etc. (Record of the Last 10 Years)	
6.1.1 Temperature - Daily Temperature (Min. and Max., in Celsius degrees) - Monthly Temperature (Mean, in Celsius degrees)	
6.1.2 Rainfall - Annual Amount of Rainfall - Annual Number of Rainy Days (Over 5 mm) - Monthly Number of Rainy Days and Rainy Season	
6.1.3 Humidity - Daily Humidity (Minimum and Maximum, %) - Monthly Humidity (Mean, %)	
6.1.4 Duration of Sunshine - Annual and Monthly Duration	
6.1.5 Limitations on Outdoor Work in Hot Weather - Annual Number of Days with Temperature over 30℃ - Annual Number of Rainy Days (Over 5 mm) - Annual Number of Foggy Days	
6.1.6 Limitations of Work in Cold Weather - Annual Number of Below-Zero Temperature Days - Annual Number of Snowy Days (Over 5 mm)	
6.2 Wind, Typhoon, Cyclone, and Tsunamis	
6.2.1 Windrose Diagram of Site	
6.2.2 Wind Velocity and Direction (Record of the Last 10 Years) - Daily Wind Velocity and Direction (Minimum and Maximum) - Monthly Mean Wind Velocity and Direction	
6.2.3 Probability of Typhoon and/or Cyclone Occurrence	
6.2.4 Annual Number of Strong Windy Days (Last 10 Years' Record) - 10 m/s, 16 m/s, and 24 m/s over	
6.2.5 Probability of Tsunami Occurrence	

Table 7 Site Investigation Checklist for Weather Conditions

A Place
Where
Ships are Built

Part II.

Designing of a New Shipyard

This is a case where the Goliath Crane's capacity in the existing dry dock was doubled and replaced in just one day, without interrupting production. This resulted in a dramatic improvement in productivity and an increase in sales.

DSME (Daewoo Shipbuilding & Marine Engineering) was acquired by Hanwha Group on December 16, 2022, and now its name was changed to Hanwha Ocean.

Part II.

Designing of a New Shipyard

Since the life cycle of products, such as ships and offshore platforms, produced by shipyards is generally very long (at least 20 to 30 years), it is almost impossible to define the life cycle of the shipbuilding industry as a country's key industry. In addition, further subdivision would make it impossible to estimate the duration of the transition process in terms of the stages of introduction, growth, maturity, and decline in the shipbuilding industry.

Therefore, when designing a new shipyard, it is not advisable to solely focus on the construction of the shipyard's overall appearance. First of all, even though the shipyard design scope is initially focused on the capacity and operation of the shipyard to manage the risk of entering the new market during the introductory stage of the industry, it is crucial to fully consider future expandability in the initial design scope. Through this, the shipyard will be able to promptly increase its production capacity by expanding its facilities during the industrial growth stage.

From this perspective, the Technological Solutions that form the design basis for a new shipyard should cover a wide range of aspects, including the following minimum scope:

Firstly, it is essential for everyone involved in this project to have a clear understanding of the modern shipbuilding processes that are currently being implemented in the world's leading shipyards.

Secondly, a vision and goal should be established based on the mid- and long-term production goals of the new shipyard. Through this process, the capacity of the new shipyard, the mid- and long-term production program, and the production manpower plan can be effectively established.

Thirdly, the production capacity required for each production process must be defined by dividing it into each shipyard's development stage, based on the mid- and long-term production program. Therefore, since the productivity of a new shipyard cannot immediately reach its highest level, it is reasonable to divide the shipyard development into at least two or three stages.

Fourthly, by systematically arranging the related production facilities at the new shipyard site, it is possible to smoothen logistics flow between each production process in the shipyard. Of course, quantitatively evaluating the optimization of the shipyard layout is not easy due to various constraints. These constraints include a variety of variables including the shape of the site, the connecting position to the outside of the shipyard boundary, and the stratum conditions. Therefore, it is desirable to gather opinions on the advantages and disadvantages of various alternatives for shipyard layout from experts in various fields and evaluate them in order of priority.

Fifthly, it is necessary to determine the basic specifications and quantity of equipment and facilities to be installed for each workshop, taking into account future expansion possibilities for each stage of shipyard development. Here, the step-by-step expansion of shipbuilding facilities should be based on the production capacity growth prospects specified in the mid- and long-term production program and the overall improvement in productivity at the shipyard.

Sixthly, establishing the infrastructure for the new shipyard's Information and Communication Technology (ICT) system means developing a shipyard operation system beyond simply installing an ICT network. It is one of the most crucial factors, and is even more important than having modern production facilities.

Seventhly, the education and training program for employees and the plan for technology transfer to establish the new shipyard's operating system must be planned in accordance with the actual operation schedule of the shipyard. Unlike other subjects, this particular section contains numerous intangible elements, which may make it difficult to plan for and implement.

In order to carry out the rational development of a new shipyard, it is necessary to establish shipyard design guidelines in at least seven key areas above. Part II covers seven topics corresponding to these guidelines and focuses on the construction of merchant ships in general. Furthermore, the corporate valuation, based on the shipyard's mid- and long-term production program, will be addressed separately in Part III.

1 General Shipbuilding Process

Although there may be variations in the production process depending on the composition of the new shipyard's business portfolio, it is essential to comprehend the shipbuilding process based on the modern Build Methods employed by the world's leading shipyards for the most prevalent merchant vessels.

1.1 Overall Shipbuilding Process

If a new ship is listed sequentially from the construction contract to delivery, it can generally be subdivided in the following order, as shown in Figure 3.

"New Shipbuilding Contract → Mid- and long-term Production Program → Basic Design → Procurement Plan → Detailed and Production Design → Material Supply → Steel Cutting and Fabrication → Block Assembly → Pre-Outfitting → Block Painting → Pre-Erection and Erection → Keel Laying → Launching → Quay Outfitting → Sea Trial → Ship Delivery"

Figure 4 provides a simplified flow of the three core value chains of the shipbuilding processes: design, procurement, and production, along with production planning. However, it is important to understand that the dependencies between predecessor and successor tasks are very complex.

Level-1 planning is a mid- and long-term production program that aligns with the shipyard's annual business plan. This planning is only finalized when project orders are received through sales activities.

In addition, level-2 planning names to baseline planning. Baseline planning involves creating a plan that serves as the foundation for design drawing issuance schedule, purchase order schedule, and production schedule for the entire year. Since this baseline planning should be established based on production technology drawings, such as block division, which serves as the foundation for the Build Method, it is crucial to specify and confirm the comprehensive Build Strategy in the initial stage of design. Level-3 and Level-4 planning are plans that correspond to the execution plan and work order for the production organization.

Figure 5 is a comprehensive reflection of the shipbuilding process depicted in the actual project schedule. According to the key event of the Suezmax class COT (Crude Oil Tanker) project, which involved the supply of drawings and materials from a Korean engineering company to a shipyard in Europe, there is a very closed correlation between the schedule for issuing drawings, the schedule for purchasing steel materials, the schedule for production technology drawings, the schedule for block assembly, and the schedule for supplying manufactured outfittings.

For reference, if this project were carried out in a Korean shipyard with a well-developed shipbuilding equipment and materials supply chain, the construction period from steel cutting to ship delivery could be shortened by approximately 11 months. Compared to shipyards in other countries, Korean shipyards have an advantage in terms of delivery time competitiveness.

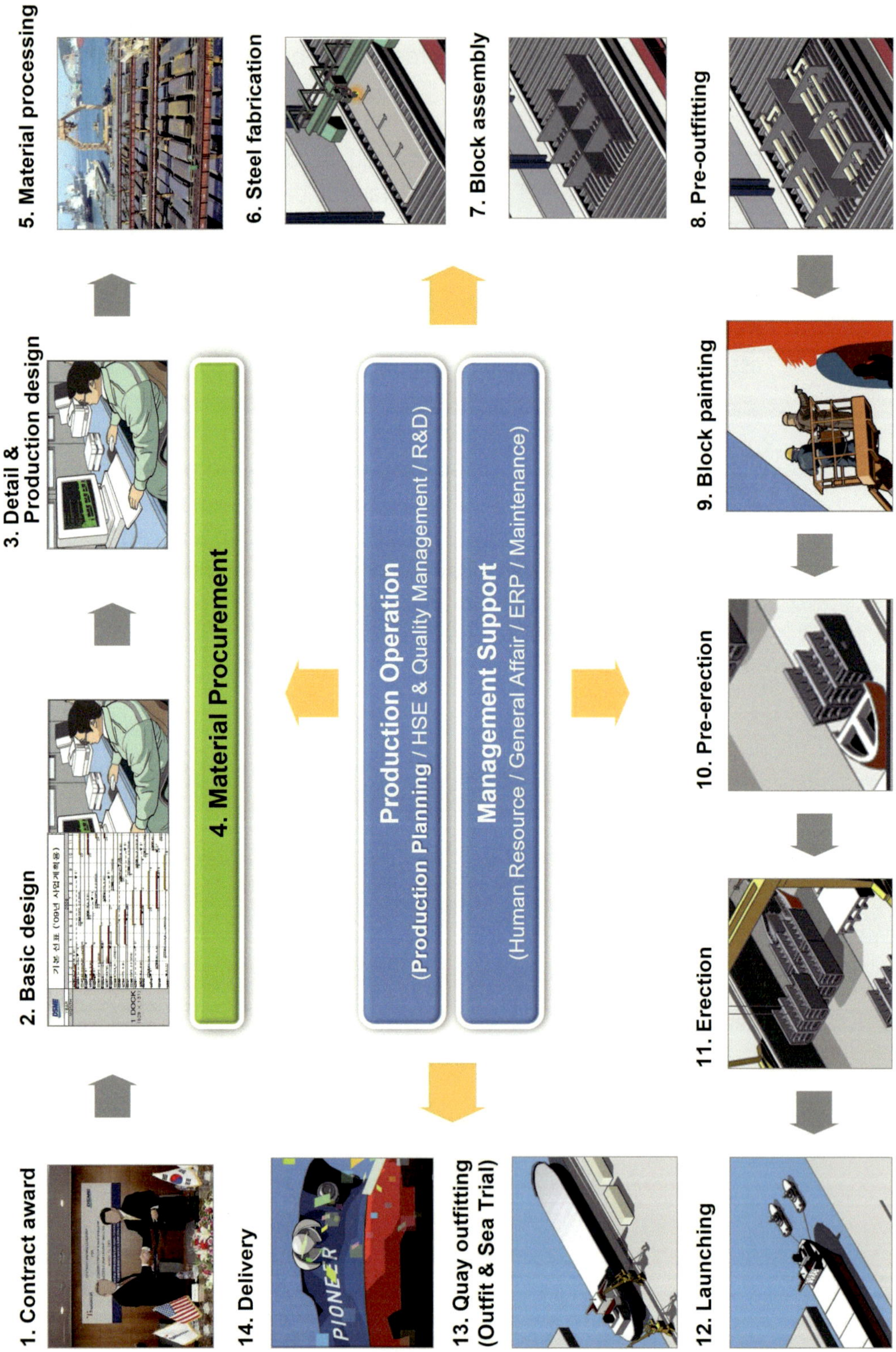

Figure 3 Overall Shipbuilding Process

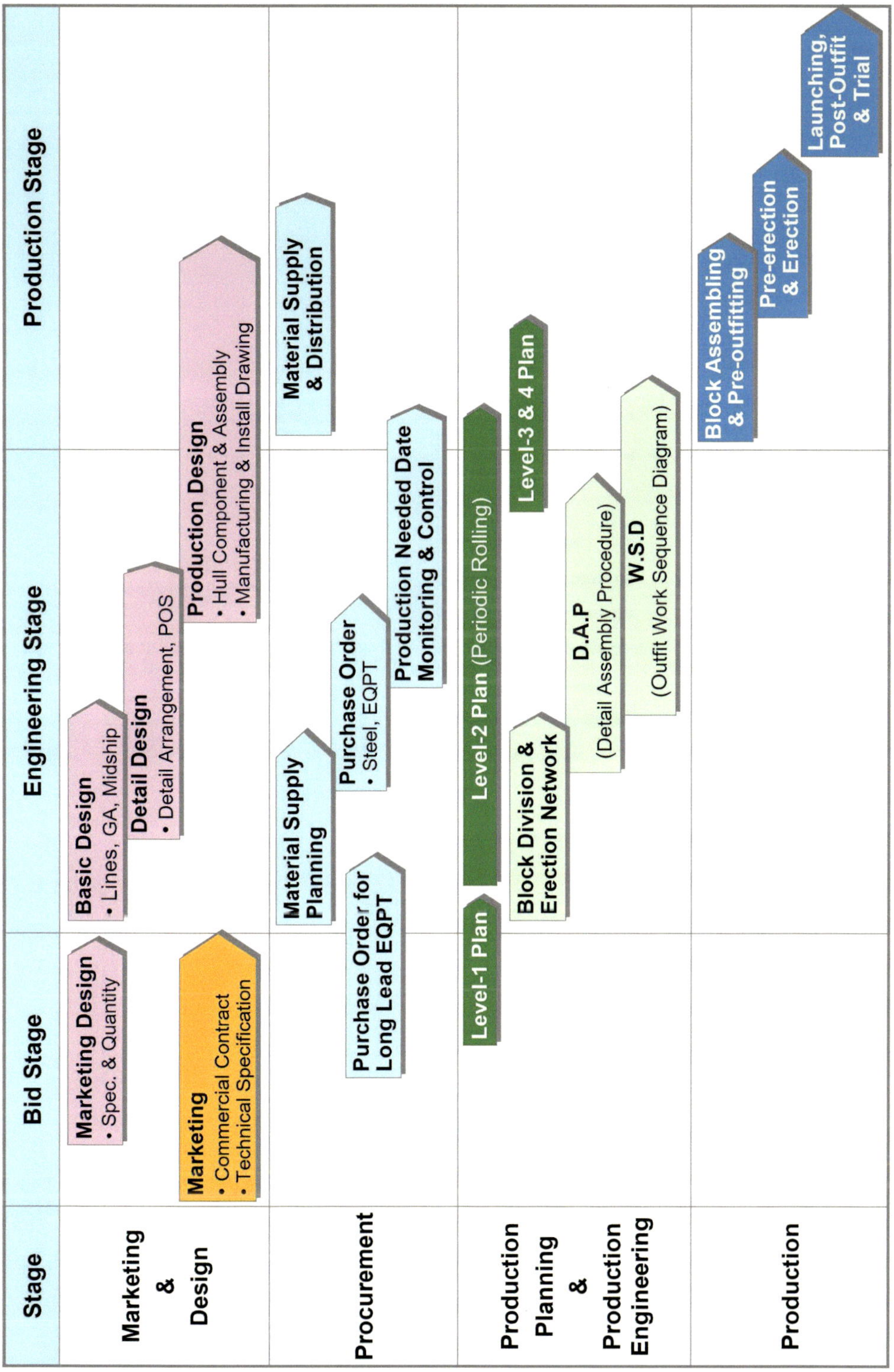

Figure 4 Sequential Shipbuilding Process

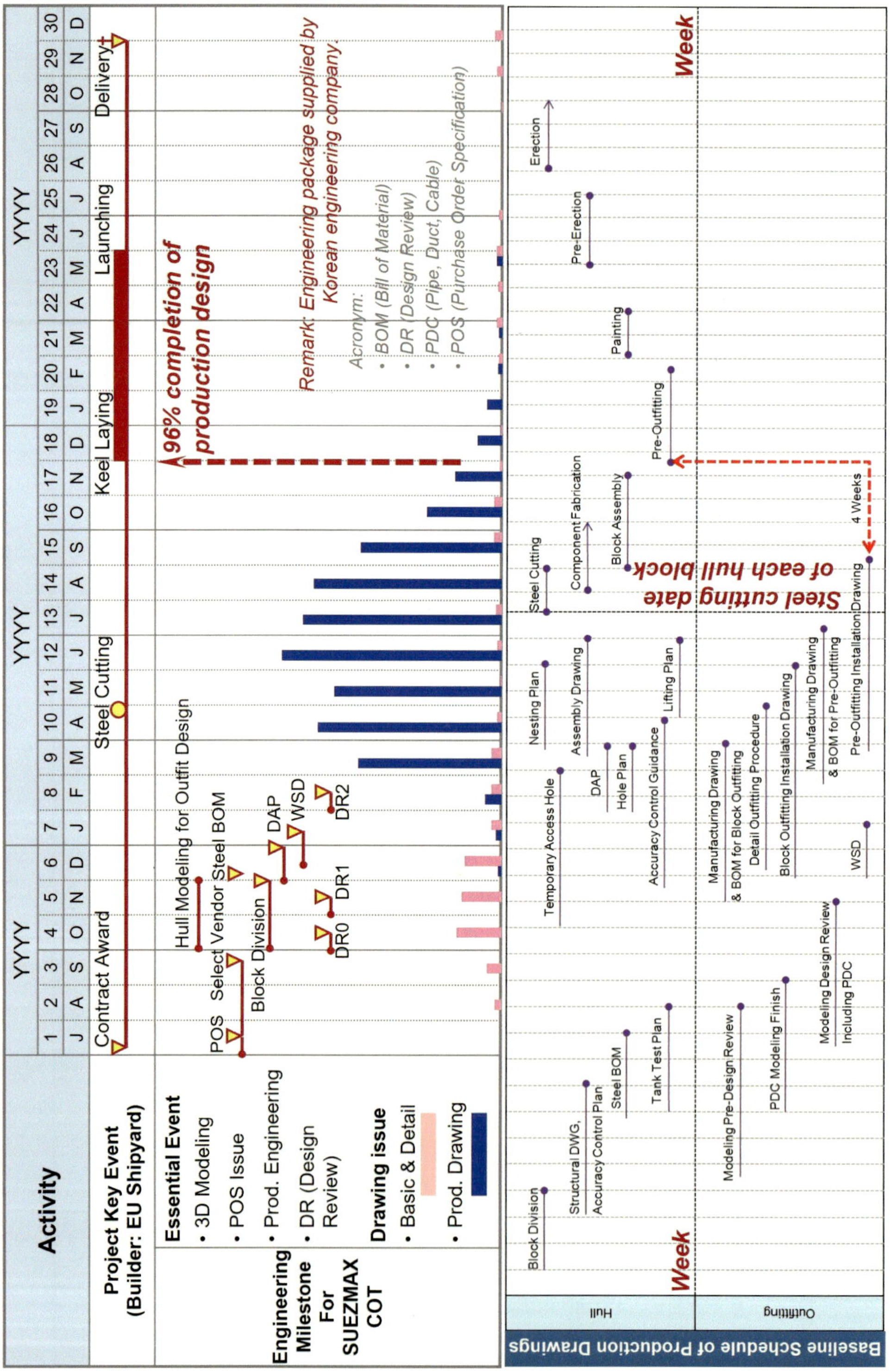

Figure 5 Correlation with Key Events and Design-Production Engineering Drawings Schedule

1.2 Production Planning Process

As mentioned earlier in the overview of the shipbuilding process, the production planning process is a critical element that plays a pivotal role in the comprehensive operating system of a shipyard. It is a coordinated process that aligns the schedules of each individual project and is directly connected to the scheduling of all functional organizations within the shipyard.

One common phenomenon in shipyards worldwide is that not all processes are executed perfectly in accordance with production planning. As a result, this issue is the most challenging field in shipyard operations. Many factors may come into play, but while there are similarities in the assembly of large structures, such as aircraft assembly, the fundamental difference in shipbuilding is that manufacturing methods are not as well standardized and automated as they are for off-the-shelf products like aircraft.

Because each shipyard has its own production planning system tailored to its specific production products, facilities, and the Build Methods, it is not easy to determine which production planning system is the most effective model. Therefore, it is desirable to regularly explore the production planning systems of top-tier shipyards worldwide and consistently enhance one's own production planning system.

In the production planning system depicted in Figure 6, Level-1 planning corresponds to the highest level of production planning established in line with a 3 to 5-year business plan. Level-2 planning is the shipyard's baseline planning, which is established based on the Level-1 planning and the dry dock's annual operating cycle. This Level-2 planning includes all production activities for six-month period and is a rolling plan that is updated quarterly. In the case of Korean shipyards, the number of dry dock operations may vary depending on their primary products, but during the boom period of the shipbuilding industry, the turnover rate of dry docks may reach up to 8 to 10 times per year.

Level-3 planning is a production schedule that is directly prepared by the production organization, based on the baseline planning. It is a two-month production schedule that is usually updated monthly. Level-4 planning is a work order in which the foreman of the unit production organization controls the work site based on Level-3 planning. This type of work order typically covers a workload that spans 2 to 4 weeks. One of the major distinctions in the operation of production planning systems between Korean shipyards and overseas shipyards is who creates the Level-3 and Level-4 planning. Unlike shipyards overseas, Korean shipyards' production departments take a strong initiative in the operation of shipyard production facilities through direct planning.

Another reason why production planning is difficult in shipyards is determining how to configure the Work Breakdown Structure (WBS), which serves as a fundamental component of all production planning. The composition of shipbuilding specifications and the design of all design disciplines primarily focus on functionality and systems. However, production activities before the ship's launch are carried out based on hull blocks and compartments. After the ship is launched, post-outfitting work progresses on systems and compartments oriented once again. As a result, since the WBS system in shipyards inevitably becomes more complex, it can have a significant impact on the quality of production planning. Therefore, special attention must be paid when developing the WBS system of the new shipyard.

The production engineering drawings depicted in Figures 7 and 5 are included in the Build Strategy and/or Build Methods at overseas shipyards. These drawings are closely linked to the establishment of production plannings. Even in Level-1 planning, block division drawings are used as crucial information to assess a shipyard's mid- and long-term workload for manpower, production facilities, and production areas.

In the case of overseas shipyards, the majority of them do not create production engineering drawings such as DAP (Detail Assembly Procedure), which are highly beneficial for hull assembly, or WSD (Work Sequence Diagram) for efficient outfitting work. In the case of Korean shipyards, in addition to micromanaging the block assembly work, the outfitting works are divided into approximately 10 stages and meticulously managed, as illustrated below. It can be said that these differences manifest as disparities in labor productivity.

On the other hand, commercial planning software, such as Primavera, is highly advanced and widely used as a project management tool nowadays. However, it appears to have limitations when used as a planning system for a shipyard operating system, rather than as its own dedicated project management tool. Therefore, based on experience, it should be noted that even with the introduction of an advanced Enterprise Resource Planning (ERP) and planning system, a significant portion of the production planning system will inevitably require customization to align with the shipyard's specific practices and needs.

[Reference] Typical working stages for outfittings in the Korean shipyard :

- *H stage : Pre-outfitting incorporated with hull assembly drawing*
- *S stage : Pre-outfitting at the component and minor assembly stage*
- *C stage : Pre-outfitting at the sub-assembly stage*
- *A stage : Pre-outfitting at the grand assembly stage*
- *B stage : Pre-outfitting at the pre-outfitting workshop stage before block painting*
- *T stage : Pre-outfitting after block turnover, but before block painting*
- *E stage : Post-outfitting after block painting, but before pre-erection or erection*
- *P stage : Post-outfitting during pre-erection stage*
- *D stage : Post-outfitting during dock stage*
- *Q stage : Post-outfitting at the quay stage after launching*

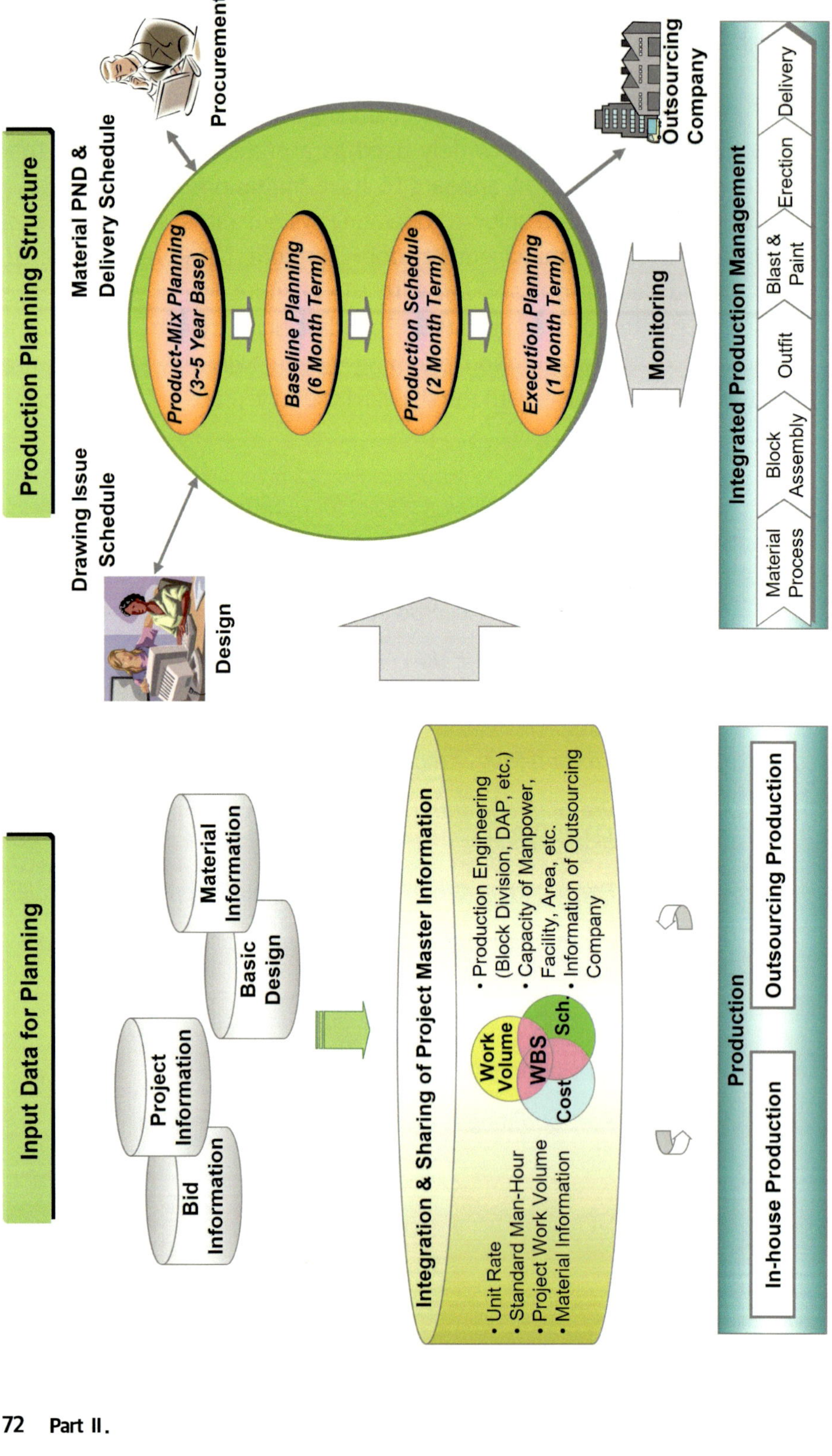

Figure 6 Production Planning Process

Description	Main Purpose	Remarks
Block Division (Integrated with level 2 & 3 planning)	**Hull division for assembly & erection capacity** • Mid & long-term workload analysis • Workload balancing • PND (Production Needed Date) assigning for material	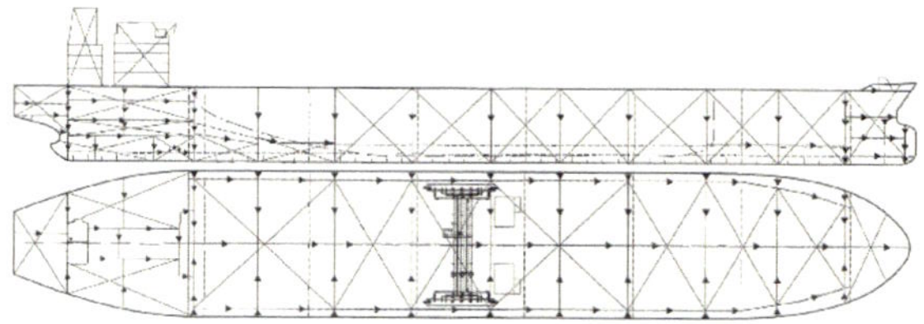
Detail Assembly Procedure (Integrated with level 3 & 4 planning)	**Assembling Procedure of hull block** • Calculation of work volume • Optimum assembly working stage	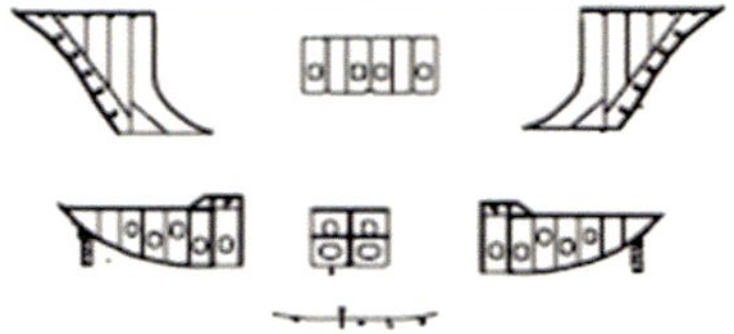
Work Sequence Diagram (Integrated with level 3 & 4 planning)	**Outfitting installation stage by block or zone** • 10 stages (in consideration of easy work & minimized work volume)	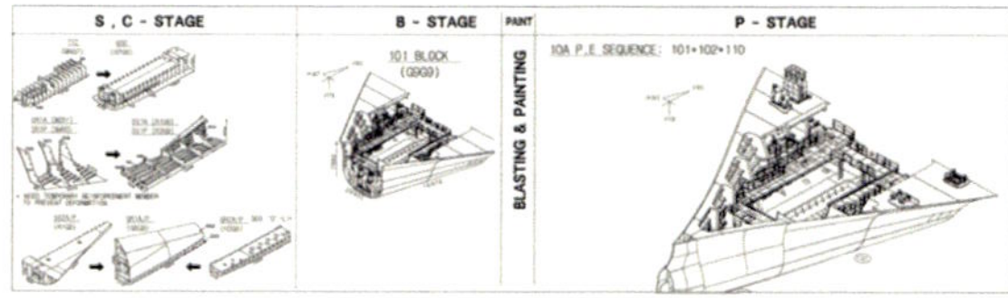

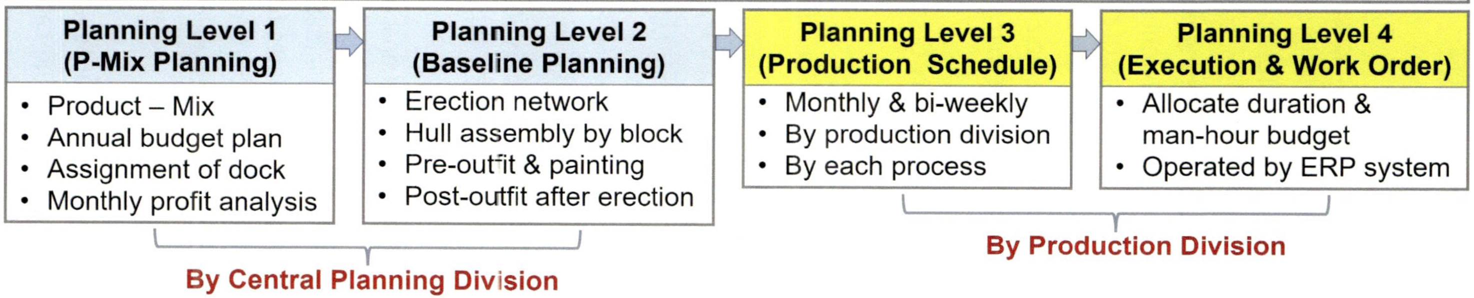

Figure 7 Planning System and Correlation with Production Engineering

1.3 Design Process

As shown in Figure 8, the ship design process is typically divided into three phases: sales engineering, basic and detailed design, and production design. In other words, design activities do not start after awarding a project order, but actually begin during the sales engineering phase, which occurs before a shipbuilding contract is signed. In this sales engineering phase, the primary design solutions that can meet the basic requirements of the shipowner, such as the principal dimensions and particulars of the ship, the price level of the ship, and the estimated delivery time, are initially developed. Based on this, the sales organization conducts the first technical clarification meeting with the shipowner to negotiate the ship price and delivery date.

The primary design solutions developed in the sales engineering phase include:

- Outline Specification
- Preliminary General Arrangement
- Preliminary Midship Section
- Preliminary Machinery Arrangement
- Preliminary Weight and Speed-Power Estimation
- Preliminary Indicative Price Estimation
- Preliminary Key Events Schedule (or Preliminary Project Master Schedule), etc.

The typical final deliverables of sales engineering include the technical shipbuilding specifications, the material list, material quantity estimates considering the material scrap ratio for ship price negotiation, and internal budget estimates for construction. In order to enhance competitiveness in delivery schedules, it is necessary to procure equipment with long lead times, such as main propulsion machinery, in advance as a condition of shipbuilding contract.

The shipyard's ability to develop new products is directly related to its basic design capability, making it an essential function for securing sustainable growth and profitability for the company. Of course, another option for carrying out a project is to receive a basic design package from a leading engineering company. However, this approach has the disadvantage of not properly reflecting the many past trials and errors that occur at the production site in the subsequent design. In addition, relying on external agencies for the design field, which is highly sensitive to construction costs, presents a contradiction in that there is a limit to controlling manufacturing costs internally. Therefore, it is desirable to have basic design capabilities in-house as soon as possible, in line with the construction of new shipyards.

The detailed and production design carried out after the basic design phase is closely linked to the shipyard's manufacturing capability. It is crucial for the design to be suitable with the shipyard's production facilities and its Build Methods in order to consistently enhance productivity. It is important to have a balanced design management that never overlooks the perspective of design for production, while also focusing on design optimization.

The importance of the following two subjects as core competencies required in the design process was fully recognized through the cooperation project for productivity innovation of overseas shipyards.

Firstly, the core capability of a shipyard stems from its functional and physical integration capability. As a comprehensive assembly industry, the core competency of shipbuilding is to effectively integrate and connect numerous systems from both functional and physical perspectives. In general, shipyards in advanced countries tend to pursue perfection in every design aspect, and this tendency can actually result in significant imbalances. Moreover, a bystander work practices that just requires each person to faithfully carry out their duties can lead to serious defects in design quality. The design process does not simply evaluate the average score, nor does it necessarily require a perfect score of 100 points. It should be a process of improving the overall level of design performance and eliminating critical defects, and this is a crucial responsibility for the project manager in the design field.

Secondly, the cost of failure should be minimized by enhancing design maturity. For example, if there is a lack of design maturity in progressing key production events, such as steel cutting and keel laying, the costs of failure will inevitably increase. Furthermore, it may also result in a significant delay in ship delivery to the shipowner. Therefore, recognizing that design maturity is a crucial aspect of corporate competitiveness, the following factors should be thoroughly considered in advance during the design process.

Securing an Appropriate Design Period

The design period, from the time the shipbuilding contract is awarded to the start of production (i.e., steel cutting date), must be reasonable. The suitability of this period should be assessed using the critical path method. In addition, close collaboration with the procurement department during the initial stage of design is one of the most important factors in obtaining design input information and parameters in a timely manner.

Concurrent Design of Hull and Outfitting

In the early stage of detailed design, it is necessary to check for physical interference between the hull structure and outfitting. Accordingly, three-dimensional modeling of the hull structure and outfitting must be performed simultaneously for the design review process before production drawings are issued.

Integration between Level-3 Planning, Production Drawing, and Production Engineering Drawing Schedules (see Figure 5)

Production drawings, purchase order specifications, and bills of materials must be issued on time in accordance with the production schedule and the materials lead time schedule. In the case of Korean shipyards, over 95% of production drawings are issued on time during keel laying, and more than 95% of Bill of Materials (BOM) are also issued on time.

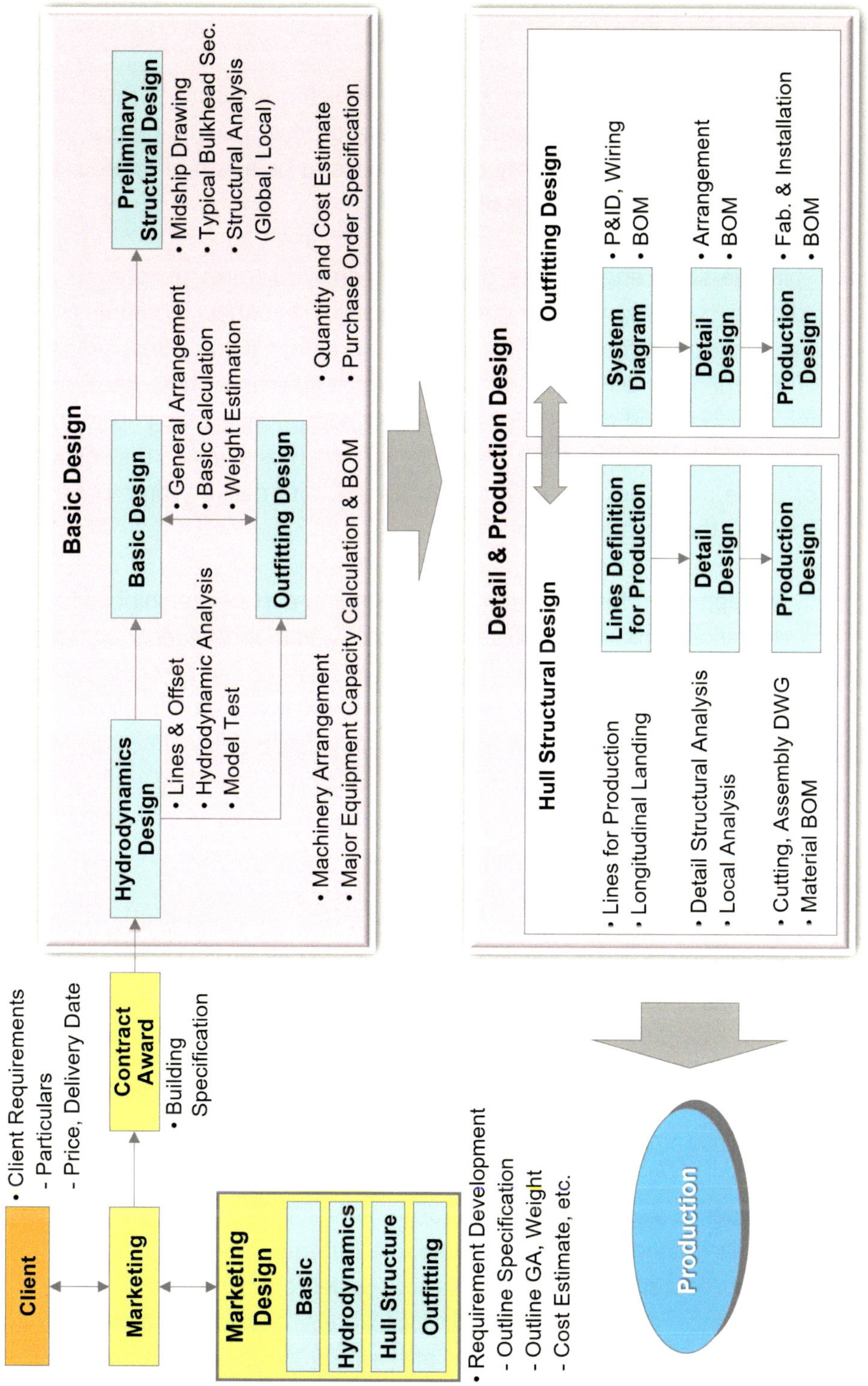

Figure 8 Design Process

1.4 Procurement Process

Looking at the procurement process depicted in Figure 9, there are no significant differences from the general product purchasing process. However, there are a few key considerations to be aware of in order to enhance the competitiveness of shipyards.

In the sales engineering phase, not only do items purchased in advance, based on the premise of a shipbuilding contract, account for a significant portion of material costs, but numerous design variables also arise depending on the results of equipment selection. These items are typically long lead time equipment, and as such, they greatly impact the overall shipbuilding timeline. Therefore, it is essential to assess the impact of these items on the construction budget and delivery time before finalizing the shipbuilding contract.

The approved Manufacturer List, which is a part of the shipbuilding specification, typically includes multiple products that have received type approval from the classification society. The final selection process takes into account the shipowner's preference and the procurement price. Due to this equipment selection process, it may cause delays in obtaining design information in a timely manner. In addition, it should be noted that emphasizing an excessively strict procedure in the name of procurement transparency may result in design delays and a decline in design maturity.

After the purchase order is placed, there are several important management activities that need to be carried out. These include monitoring and controlling the manufacturer's drawing approval schedule, factory acceptance test schedule, shipment schedule, Production Needed Date (PND) schedule, and manufacturer's service engineer schedule. In particular, when dealing with monopoly or oligopoly items, it is necessary to pay close attention to the management of these schedules.

In order to maintain a reliable procurement supply chain, it is essential to regularly conduct qualification surveys for equipment manufacturers. These surveys should assess factors such as delivery performance, quality performance, and corporate credit evaluation. These procedures are in accordance with the requirements of the ISO9001 Quality Management

System of the International Organization for Standardization (ISO).

It is desirable to simplify procurement tasks by maintaining an appropriate inventory level calculated based on consumption during a certain production period for standard materials such as steel sections and profiles, piping materials, and fastening materials commonly required in many projects.

Three essential elements for starting production are the preparation of production manpower, drawings, and materials in advance. From this perspective, it is important to establish an internal material distribution system in the shipyard. This system should ensure that the materials required to fulfill work orders can be distributed in a timely manner to the relevant production organization. It should be noted that neglecting the importance of this material distribution system or working on multiple projects simultaneously can lead to considerable confusion or material waste, even after the materials have been delivered to the shipyard.

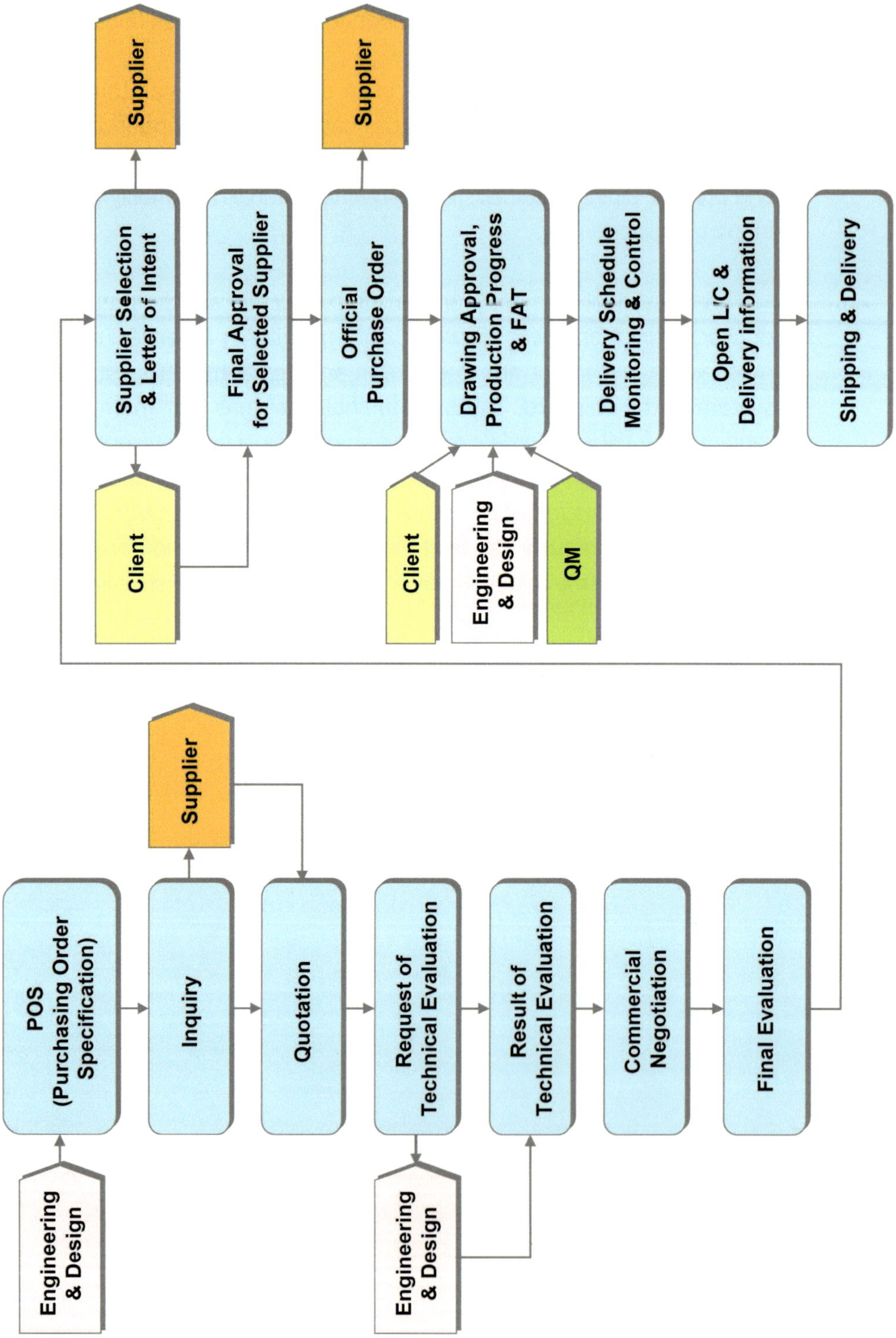

Figure 9 Procurement Process

1.5 Production Process

From now on, each production process will be briefly described, along with reference photos, in the order of shipbuilding. This will include the receipt of materials, such as steel plates and equipment, all the way to the delivery of the ship.

Materials Unloading and Steel Stock

Depending on the nature of the equipment or materials required for shipbuilding, they are delivered to the shipyard by land or sea transportation.

Steel plates, which are a common raw material used in shipbuilding, can typically be easily transported by road or rail if they meet standardized steel specifications produced at a steel mill. However, in order to enhance the productivity of shipyards, steel plates that exceed standard dimensions in length and width are transported by sea using bulk carriers, barges, or dedicated carriers.

[Unloading Crane & Steel Stock Yard]

In general, steel materials are unloaded using a dedicated magnetic crane or transporter at the unloading quay and then transported to the steel stock yard.

The storage location and order of steel plates in the

[Unloading Heavy Equipment]

steel stock yard are determined by the volume balance of each bay and the order in which they are input into the steel treatment process. With recent advancements in ICT systems, steel stock yards can now be operated using unmanned and automated operations.

In the case of machinery and equipment, they are typically stored in appropriately dispersed warehouses within the shipyard until they are ready to be installed at the production site, in accordance with the shipbuilding process. Large and heavy equipment, such as main propulsion engines and generators, is delivered just before installation and is directly erected onto ships using the Goliath crane in the dry dock or installed in the pre-erection area.

Steel Treatment

Steel plates and profiles transferred from the steel stock yard are typically moved using a conveyor system to the steel treatment process. This process consists of a series of continuous steps, starting with the removal of rust and foreign substances and ending with shop primer and drying, as described below.

[Blasting, Spraying Primer & Drying]

- Heating :
  Foreign substances and moisture are removed from the surface of the steel.
- Blasting :
  Foreign substances and rust are removed.
- Spraying :
  Shop primer coating is applied during the shipbuilding process to prevent rust.

- Drying:

  The painted surface is dried instantly using a dryer.

Afterward, the steel materials are transferred to the steel cutting process through the conveyor system.

Steel Cutting

Shot-blasted steel plates are put into various types of cutting machines for efficient cutting, depending on their intended purpose. First, after the steel plates and sections are transferred from the steel treatment workshop to the transfer bay of the cutting workshop, they are sorted by the cutting machine to be input by an auto-transfer equipment consisting of a line conveyor, cross conveyor, and sensor. And then, they are moved to the cutting machine in the designated cutting bay using overhead cranes.

[NC Plasma Cutting]

There are various types of cutting machines, such as numerical control plasma cutting, submerged plasma cutting, gas cutting, and hand cutting. However, in the case of submerged plasma cutting, which is primarily used to control cutting thermal deformation, it is not widely employed except for cutting very thin plates due to its low productivity.

In the case of numerical control strip cutting machines used for producing T-Bar shaped steel, they are equipped with multiple cutting torches to enhance the productivity of the cutting machine.

In addition, the identification number is automatically printed on the cut piece by an automatic marking device, eliminating the need for manual work as it was done in the past. Afterward, the cut pieces of various shapes are palletized for each subsequent process. They are then either directly put into the assembly line or moved to a temporary buffer area around the cutting workshop.

On the other hand, the stability of the transfer speed and gas pressure of the cutting machine is crucial as it directly impacts the quality of the cut and the subsequent welding quality.

The steel cutting process signifies the beginning of ship production, and the first day of steel cutting for the project is also marked with an official event where prayers for safety are offered. Furthermore, this date is one of the first key events in the contract related to the payment terms of the ship price.

[Various Type of Press Machines]

Steel Fabrication

Except for flat plates, there are various steel fabrication processes depending on the shape of the steel piece prior to assembly. In this process, various types of hydraulic machines are used for cold bending and line heating to make three-dimensional shaping.

In the case of a two-dimensional curved plate, it is bent using a large plate roll press and bending machine. For a three-dimensional curved shell plate, a line heating and water-cooling process is added to achieve the desired shape. Afterward, the accuracy is checked using a wooden template. In add-ition, the pin-type portal press machine and line heating are used to create the 3-dimensional shape of relatively small and thick plates located at the fore and aft of the ship.

[Line Heating for Making Side Shell Plate]

The corrugated bulkhead or hopper structure in the cargo hold is formed into a knuckle shape using a knuckle press machine.

According to the shape of the faceplate among the internal members of the hull structure, round-type members are fabricated using the horizontal roll press machine, while knuckle-type members are fabricated using the horizontal pin press machine. Flange-type members are not shaped by welding, but a flang-ing machine is used to increase productivity.

T-Bar Manufacturing

In general, bulb plates are commonly used as longitudinal members or stiffeners in many countries because of their superior stiffness-to-weight ratio. However, there are some difficulties, such as the increased probability of welding defects due to multi-layer welding in the bulb-shaped part, and the limitation of partial twist of the bulb plate

[T-Bar Manufacturing]

at the point where the curvature of the shell plate changes. Therefore, Korean shipyards manufacture and use T-Bars instead of bulb plates unless there is a specific request from the shipowner.

The main task in the T-Bar workshop is to cut the steel plate using a strip cutting machine. This steel pieces will be used as the web plate and face plate. After the cut pieces are transferred to the auto-setup device by an overhead crane, the web plate is automatically positioned in a T-shape on the face plate by the device. Then, it is transferred by a conveyor line and automatically welded by an auto-tack welding device. The tack-welded T-Bar is transferred to the welding conveyor line by an overhead crane, and then automatic full welding is performed.

The welded T-Bar undergoes a quality inspection for welding and is then transferred to a conveyor line for straightening using an overhead crane. This process straightens the T-Bar, which may have become deformed due to the heat from welding, into a straight line.

After compressing the edge of the face plate with a roller, blasting and painting work is performed on the welded section of the T-Bar web and face plate.

The bundles of manufactured T-Bar are palletized on a dedicated T-Bar pallet in the adjacent area and then moved into the subsequent assembly process.

Block Assembling

Depending on the size, weight, and shape of the hull assembly, it is generally classified as follows.
However, it should be noted that in some cases, the names used by each shipyard may be slightly different.

[Component & Minor Assembly]

- Component or minor assembly
- Sub-assembly
- Grand assembly

In the component or minor assembly workshop, relatively small-sized pieces are assem-bled into a plane shape, which is usually required regardless of the shape of the hull block. The sub-assembly process involves assembling multiple minor assemblies in a bay next to a grand assembly workshop to create a three-dimensional structure. In general, there are three types of grand assembly workshops:panel block workshop, deckhouse block workshop, and curved block workshop. The type of workshop depends on the shape of the final block.

Panel Block

The panel block workshop is a factory that subdivides and operates the assembly line based on the tact time for flat-type blocks, which make up 55 to 70% of the total block weight. Although the layout of this panel block workshop may differ between shipyards, it generally consists of seven assembly areas, as follows:

- **Bay 1**: This is a working area where primary plates are cut using a cutting machine that is capable of beveling. After the cut plates are positioned with a positioner, tack welding is performed between the primary plates using a fit-up machine.

[Sub-Assembly]

- **Bay 2**: The tack-welded welding line is welded on one side of the primary plate using the Submerged Arc Welding (SAW) technique.

- **Bay 3**: To weld the back side of the primary plates, the primary plates that has been welded on one side is flipped over using an overhead crane or a dedicated rotation device. Then, the back side is also welded using the SAW technique. The positions where the longitudinal and transverse members will be attached are marked automatically after conducting a non-destructive test on the welding line.

[Panel Block Assembling]

- **Bay 4**: After positioning and tack welding longitudinal members such as T-Bar or angle on the primary plates with the fit-up machine, full welding is immediately performed using the welding gantry. Depending on the annual production target of the panel block, it may be arranged in pairs from Bay 4 to the final assembly line.

- **Bay 5**: In this bay, the pull-to-fit method (also known as the Slit Method) is used to accurately position the transverse assembly. This is achieved by using an overhead crane and a pulling winch. As a result, it can greatly contribute to productivity improvement, as assembly is possible without the need for separate color plate work to be attached to the transverse and longitudinal members. Adoption of this Build Method is only possible with excellent accuracy control capabilities.

- **Bay 6**: Bay 6 is the re-entry working area for blocks that were partially assembled from Bay 7 in order to complete the double hull structure, including the double bottom and double side shell. After using two large-capacity overhead cranes to turn over the block, the same pull-to-fit method is utilized again as in the previous bay. The block is then placed on the assembly of the other structure that was transferred from Bay 5. Only the necessary amount of welding is performed to meet the tact time of the panel block workshop before it is moved to the next working bay.

- **Bay 7**: Hull structuralized outfittings are installed in parallel with the finishing welding of the single hull structure or double hull structure transferred from Bay 6. Afterward, the completed block is taken out to the transport carriage and moved to the block inspection area, which is located near the panel block workshop. In this area, finishing work such as grinding is carried out, and the block is inspected by shipowners and supervisors from the classification society.

Deckhouse Block

The deckhouse block is a flat block primarily designated for accommodation spaces, comprising only 3% to 5% of the total block weight. Therefore, this assembly workshop is relatively small in size and has simple production facilities. When different types of ships with thin-plate structures, such as Ro-Ro ships, passenger ships, and naval and coast guard ships, are not constructed simultaneously, there is a significant fluctuation in the workload of this assembly workshop. Most large Korean shipyards outsource the assembly of this type of block to external companies instead of handling it in-house.

On the other hand, since it has a relatively thin plate structure, controlling thermal deformation during the assembly work is one of the key factors in ensuring quality.

Curved Block

The proportion of curved blocks may be higher in the case of high-speed ships, such as container ships and passenger ships. However, for general merchant cargo ships, the fore and aft hull structure accounts for 30 to 40% of the total weight of the ship's hull block. The workshop that assembles these curved shaping blocks typically consists of three working bays.

- **Bay1**: This bay is a workspace for fabricating minor assemblies.

- **Bay2**: This bay is equipped with height-adjustable pin jigs and allows workers to assemble curved shell plates and structural members into sub-assemblies in a top-down posture. Some shipyards install line production facilities in this bay, and these facilities are installed on turntables. However, since line production generally has varying work lead times for each stage, depending on the type of block or ship, there may be certain limitations to ensure smooth line operation.

- **Bay 3**: The final assembly of the sub-assemblies supplied from Bays 1 and 2 into a grand assembly block is performed on a pin jig. During this process, the installation of hull structural equipment is also conducted.

In particular, the overhead crane in this workshop must have sufficient capacity to prevent the splitting of the main propulsion engine foundation block.

[Curved Block Assembling]

As with panel block assembly, the final curved block is transported to the block inspection area, where it undergoes finishing work like grinding, and is subsequently inspected by ship owners and classification societies.

Pre-Outfitting

Excluding the block joint part, all welding work on outfittings and the installation of equipment, piping, ducts, and electrical wires must be completed before the painting work begins. As mentioned above, the pre-outfitting work must be managed by dividing the sequence of pre-outfitting works into detailed tasks. This will allow for easier execution in a comfortable working environment and posture.

The pre-outfitting workshop is situated along the route connecting the assembly workshop and the painting workshop. By consolidating and modularizing the outfittings in the equipment concentration area, such as the engine room, productivity can be greatly enhanced. In the case of large-sized unitized outfittings, they may also be installed in the pre-erection area or dry dock.

[Pre-Outfitting & Unit Outfitting]

Before the blasting and painting process, it is necessary to carry out preservation and clogging activities to prevent contaminants from entering the equipment and pipes that are installed during the pre-outfitting process.

On the other hand, in countries with a well-developed shipbuilding industry infrastructure, such as Korea, most of the manufactured outfittings are procured through outsourcing. However, in some cases, it may be necessary to establish workshops for pipe manufacturing, hull outfittings manufacturing, carpentry, and galvanizing in the shipyard. In particular, it is necessary to have the capability to manufacture piping in a suitable capacity at the shipyard. This is especially important when there are numerous offshore projects that require very thick steel piping or non-ferrous piping, as well as the fabrication of on-site adapted pipes for erection joint parts.

In the case of galvanizing workshop for steel piping and pickling workshop for non-ferrous piping, it is essential to strictly observe relevant regulations to minimize risks to both humans and the environment.

Blasting and Painting

Even steel materials that have been undercoated with shop primer may still experience corrosion due to welding spatter, weld contamination, scratches, and other factors during block assembly.

Therefore, high-pressure blasting with grit or steel balls is performed before block painting. In particular, a dehumidifier must be installed in the blasting workshop to prevent immediate corrosion in the bare steel conditions after blasting.

Since painting is typically done 2 to 4 times, depending on the required film thickness specified in the shipbuilding specification, it is important to take into account the necessary drying time between each painting interval. According to the climatic conditions of the shipyard, the ratio of the number of workshops in the blasting workshop to the painting workshop should be appropriately considered. In addition, if the shipyard's climatic conditions are unfavorable, the painting process can be configured as a line production system equipped with painting and drying facilities following the blasting process.

[Blasting & Painting]

As this process is entirely dependent on manpower, it is important to wear Personal Protective Equipment (PPE) properly. Additionally, due to the hazardous nature of the working environment, which includes factors such as large movement, noise, and limited visibility, it is crucial to exercise caution, particularly when installing scaffolding with stanchions and handrails.

Pre-Erection

To reduce the amount of hull block erections in the dry dock, painted hull blocks are erected in advance in the designated pre-erection area adjacent to the dry dock or in the land-based construction area. This pre-erection is done in accordance with the erection network diagram, with the exception of certain individual erection blocks. The size of the pre-erection block is determined based on the safe working load of one or two large gantry cranes, also known as Goliath cranes, in the dry dock. It is desirable to provide that the required area for pre-erection is at least as large as the floor area occupied by the launching vessel in the dry dock.

[Pre-Erection]

Additionally, pre-outfitting works, which were suspended at the block joint parts, are carried out simultaneously, and large unitizing outfitting is also installed if there is.

Erection

Keel laying, which involves placing the first block on the dry dock, is the second key event in the shipbuilding process. In general, the double bottom block of the engine room, which corresponds to the critical path that affects the duration of the dry dock period, is placed on the dry dock as the keel laying block. After that, the erection progresses in the bow, stern, and upward directions, simultaneously, following the erection network.

Recently, some Korean shipyards have begun using a new Build Method of lowering pre-erection blocks weighing more than 3,000 tons directly from in front of the dock gate to the bottom of the dry dock using a high-capacity floating crane. These blocks are then transported to their designated positions by multiple synchronizing transporters. By adopting this new Build Method, the number of annual dry dock operations can be increased, along with production volume and sales.

[Keel Laying & Erection]

Depending on the size of the dry dock, it is possible to build four ships simultaneously using a tandem and parallel construction method in a dry dock. The specific details to be considered when designing and determining the dimensions of the dry dock will be discussed later.

- Tandem Construction:
  The full-ship is built in the dock gate direction, and the half-ship is built in the dock head direction.
- Parallel Construction:
  Ships are built in two rows, one on the left side and one on the right side of the dry dock.

In the case of shipbuilding in a land-based construction area, the shipbuilding process is almost the same as that in a dry dock. However, a separate transfer system, such as the gripper jack system and launching barge, is required to move and launch the ship.

In the meantime, there have been recent cases of constructing very large container ships of 20,000 TEU or more in a floating dock. In such cases, the use of a floating crane for mega-block erection purposes is essential. Additionally, a high level of expertise is required for operating the floating dock and ensuring quality of accuracy control.

During the dry dock period, various tasks are also carried out. These include the installation of the main propulsion engine and shaft system, as well as the machining of the propeller shaft and the contact machining for the propeller and rudder, respectively.

[New Shipbuilding on Floating Dock]

Tank hydrostatic tests are conducted to assess the strength of the tank's hull structure.
Additionally, shore power connection tests for on board tasks at the quay-side after the ship's launching are performed before ships launch.

The last step in the dock process before launching is to paint the bottom and side shell as a final painting. The final painting on the area where the keel blocks were positioned is done by moving the keel blocks one by one.

[Propeller Shaft Machining & Installation]

Launching

Launching is the third key event under the shipbuilding contract, and the ship is floated when seawater is flowed into the dry dock. Next, the dock gate is opened, and the vessel is moved to the mooring quay using tugboats. In order to prevent excessive trim by stern or heeling during

the ship floating process, it is necessary to check the ship's floating condition before launching. In addition, in tandem construction, where there is no intermediate gate in the dry dock, the half-ship is re-docked to the location of the ship in the next batch of dry dock.
This tandem construction is a highly useful shipbuilding method for evenly distributing the workload in the dry dock.

[Launching @ Dry Dock, Floating Dock & Launching Barge]

The launching process at the land-based construction area is typically conducted in a shore area with adequate water depth after the ship has been transferred to the launching barge. In order to prevent damage to the hull bottom caused by vertical movement during the floating process, it is crucial to control the area and ensure that no vessels are operating nearby prior to the launch.

[Load-out after Land Based Construction]

On the other hand, in the case of shipbuilding in a floating dock, the launching process is usually carried out in its position.

Quay Outfitting (Post-Outfitting)

Until the official sea trial, the vessel is moored at the quay and undergoing post-outfitting work. Depending on the type of ship, the duration of this quay period can take less than two months for general cargo ships, while others, such as LNG carriers or drill ships, typically require more than 10 months.

When the productivity of a shipyard increases, the time spent at the quay naturally decreases. However, on the other hand, as the number of ships being built increases, the workload at the quay may become a bottleneck process. Therefore, it is necessary to consider future expandability in advance when designing the shipyard layout.

[Quay Outfitting & Dock Trial]

Major post-outfitting works include the installation of remaining equipment, such as accommodation facilities, wiring connections for electrical equipment, and repair painting work. In addition, before the official sea trial, various tests are conducted, including tank tests, test runs of the main propulsion engine and generator, lifeboat drop tests, fire extinguishing facility tests, and more. Dock trials are also carried out to ensure the proper functioning of all safety and navigation devices.

Furthermore, an inclining experiment is performed to verify the lightship weight, vertical center of gravity, and deadweight just before the official sea trial.

In shipyards located in areas prone to tropical storms, such as typhoons, it is essential to conduct mooring analysis to ensure the strength of quay mooring devices and mooring ropes.

Official Sea Trial

When the post-outfitting work and dock trial are completed at the quayside, the ship moves to the designated sea trial area and conducts official trials at sea to verify the ship's performance.

In the case of a ship, such as an LNG carrier, that spends a long period of time at the quayside, it is necessary to redock the ship at the floating dock before conducting sea trials. This is done to remove marine growth, such as moss, barnacles, and mussels, that accumulates on the underwater surface.

[Sea Trial]

Regardless of the type of ship, the common test items for sea trials include the main engine starting test, speed trial, endurance test, Unmanned Machinery Space (UMS) test, fuel consumption test, steering gear test, maneuvering trial, noise and vibration test, mooring and anchoring test, etc. All these tests are conducted in the presence of the shipowners and the surveyors from the classification society.

If the basic performance, which includes a performance guarantee clause in the contract, falls outside the permissible range, the contracting parties engage in sensitive negotiations regarding the liquidated damages based on the test results at the time of delivery.

Naming Ceremony and Delivery

When the sea trial is successfully completed, a grand naming ceremony is held in the presence of VIPs from the shipowner, classification societies, and the shipyard. Additionally, various certificates are issued by classification societies and government authorities. When the punch list work is completed and all marine stores, such as oil, food, and charts, are loaded, the vessel departs for its scheduled port of call.

At the time of departure, after the ship is delivered, a guarantee engineer from the shipyard accompanies the ship to prepare for any potential failures, such as unexpected unskilled operation or malfunctions, during its first voyage.

[Naming Ceremony]

So far, the modern shipbuilding process in the world's leading shipyards has been examined. Figure 10 provides a concise, diagrammatic overview of the major shipbuilding processes.

"Modern Shipbuilding Processes," a 3D graphic animation, is provided *as Appendix A* to enhance understanding of the shipbuilding process. For reference, this material contains general information provided by DSME in Korea for training purposes at Russian partner shipyards in the past.

Figure 10 Production Process

2 Production Goal of a New Shipyard

As mentioned earlier, the shipbuilding industry is a representative industry that relies heavily on manpower and production facilities. It is not feasible to immediately achieve the desired productivity or output solely by having a modern shipyard facility. There may be numerous factors, but it inevitably takes a considerable amount of time to achieve the production goal. Therefore, it is necessary to develop the production goal gradually, step by step. Therefore, first of all, an annual target for project awards for new shipyards is set based on an analysis of productivity in existing shipyards and market research conducted in the country. Based on this, it is desirable to establish the demand for resources such as the size, production capacity, and required manpower of the new shipyard.

2.1 Production Goals on an Annual Basis

Using the format of Table 1, which is the market research summary table presented as an example at the beginning of this book, it is necessary to first determine the type of ship, class of ship, and number of ships to be built for a minimum of 10 years. Furthermore, based on the production volume for these types of ships, the following representative annual production goals for new shipyards are set, as shown in Figure 11.

- Production capacity by ship type (Number of ships)
- Steel treatment tonnage (tons) and hull assembly weight (tons)
- Required manpower (persons)

In order to establish stretch goals for a new shipyard, long-term production programs must be set as part of resource planning. To establish an efficient production program, it is crucial to identify the bottleneck resource in all assumptions and undergo multiple rounds of refining and optimizing the process.

Key assumption include :

- Product mix plan

- Key events date (such as contract award, steel cutting, keel laying, launching, and delivery)
- Production capacity plan (such as the number and size of docks, quay length, etc.)
- Standard information by ship type (such as construction period, steel weight, scrap ratio of steel, production volume, production man-hours, and standard S-curve.)
- Annual target of productivity improvement (such as construction duration and man-hours)

Typical key outputs include :

- Mid- and long-term production program (in graphical form)
- Dock operation plan (such as the number of dock turnovers, dock batch plan)
- Key events schedule by project
- Annual sales
- In-house production capacity (such as steel treatment, assembly weight, lead time at the quay, pre-erection area, and production manpower) as well as outsourcing demand
- Bottleneck processes and new investment demands

The production program of a shipyard is the highest level of production plan. It plays a crucial role in reviewing and adjusting the production workload for a period of generally 3 to 5 years, taking into consideration the time from contract to delivery of the ship. However, when it comes to establishing a new shipyard, it takes a significant amount of time to develop an operating system and achieve a certain level of productivity. Therefore, it is necessary to set long-term goals for at least 10 years. Since one of the most significant factors in this analysis is determining productivity assumptions at the start of shipyard operations and developing productivity improvement plans over time, it is crucial to give special attention and conduct objective evaluation. Therefore, dividing the shipyard facility development stage into two or more stages, if possible, can be a more reasonable approach to expanding shipyard production capacity. This approach can help compensate for the numerous variables involved.

Number of Vessels (Annual)
- **Total : 8 ~ 10 Vessels**
- COT : 5 ~ 6 Vessels
- LNGC : 3 ~ 4 Vessels

Assembly Weight (Annual)
- **Total : 200,000 tons**
- Vessel : 185,000 tons
- Offshore : 15,000 tons

Number of Employee (Maximum)
- **Total : 7,300 p**
- Worker : 5,850 p
- Manager & Engineer : 1,450 p

Categorization		1st Year	2nd Year	3rd Year	4th Year	5th Yea	6th Year	7th Year	8th Year	9th Year	10th Year
Number of Vessel (Launching Basis)	COT (Aframax Class)	-	-	1	4	5	2	1	5	5	4
	LNGC (174K)	-	-	-	-		3	3	4	4	4
	Offshore (FPU)	-	-	-	-	-	-	-	-	-	1
	Sub-Total	**-**	**-**	**1**	**4**	**5**	**5**	**4**	**9**	**9**	**9**
Steel Processing Weight (ton)		**-**	**556**	**75,498**	**79,342**	**72,930**	**110,981**	**162,100**	**185,283**	**206,404**	**215,094**
Structure Assembly Weight (ton)	COT	-	473	38,695	67,440	59,165	1,093	64,636	73,913	70,058	65,116
	LNGC	-	-	-	-	2,826	44,763	73,149	83,577	83,946	52,266
	Offshore	-	-	25,478	-	-	48,478	-	-	21,440	82,447
	Sub-Total	**-**	**473**	**64,173**	**67,440**	**61,990**	**94,334**	**137,785**	**157,490**	**175,444**	**199,830**
Required Manpower (person)	Worker (Dir. + Indir.)	Recruit & Training	-	1,896	1,752	1,718	2,573	3,768	4,474	5,834	5,941
	Manager & Engineer		-	425	425	425	562	827	1,089	1,408	1,433
	Executive		-	23	25	25	37	40	40	40	40
	Sub-Total		**-**	**2,344**	**2,202**	**2,168**	**3,172**	**4,635**	**5,603**	**7,282**	**7,414**

Figure 11 Production Goals for 10-Year Operation *(Example)*

2.2 Productivity Improvement Target

It is natural to apply rationality when setting target values, even for challenging production goals. However, comparing and evaluating based on objective data is not easy because most developing countries lack such data for their shipbuilding industries. If defining one's own level of productivity is not feasible, an alternative is to utilize data from neighboring countries that have similar economic systems or working cultures. After that, it will be possible to achieve a certain level of objectivity in determining the target value of productivity by comparing it to typical values in countries with well-developed shipbuilding industries, such as Korea.

As indicators for comparing and evaluating labor productivity in the shipbuilding sector, the construction period and production man-hours by ship type are typical productivity comparison metrics. It should be noted that the figures referenced in Table 8 represent values from a time when labor productivity in Korea was relatively excellent.

Categorization		Shipbuilding Period (Month)				M-Hour/T on
		Total	SC~KL	KL~LC	LC~DL	
Korean Shipyard	COT	8.5~10.5	4.5~5.5	2.5~3.0	1.5~2.0	13~15
	LNGC	17.0~18.0	4.5~5.5	2.5~3.0	9.0~10.0	38~40

** Note: SC - Steel Cutting, KL - Keel Laying, LC - Launching, DL - Delivery*

Table 8 Comparison of Productivity (Referenced Figures in Korea)

There are three key factors that directly impact shipyard productivity indicators: labor force, production facilities, and shipyard operating system. Table 9 shows the results of estimating the sensitivity of each factor's effect on production man-hours, based on shipbuilding performance.

Description	Total	Workforce	Facilities	Operating System
Sensitivity of Productivity	100.0%	72.0%	13.0%	15.0%

Table 9 Sensitivity Analysis of Productivity

First of all, it shows an important implication that productivity is greatly influenced by the skills cultivated through education and training of the workforces, as well as their perception of labor. And, while it may not seem like a big deal, the operating system of the shipyard has a significant impact on productivity, even more so than modern production facilities. Of course, these figures are based on an analysis of Korean shipyards, but if a shipyard's operating system is not properly developed, productivity will fall short of expectations no matter how state-of-the-art its production facilities are. Since these cases can be easily found, it cannot be overstated how important it is to establish a proper operating system in a new shipyard.

Regarding the productivity of the labor force, the learning effect of the new shipyard's accumulated shipbuilding experience is significant and cannot be ignored. The learning curve depicted in Figure 12 demonstrates the performance data of a European shipyard that has received technology transfer from a Korean shipyard and shows how much it can contribute to improving productivity. Therefore, when setting target values for the construction period and production man-hours of a new shipyard, it is important to fully consider this learning effect.

Climatic conditions cannot be overlooked, as they are among the factors that affect productivity. Except for the small size of ship construction, outdoor work is generally unavoidable after the pre-erection process. Therefore, outdoor conditions inevitably have a direct impact on productivity due to their effects on welding, painting, electrical work, traffic, and worker health factors. In regions with moderately extreme climatic conditions, the production program must be developed by subtracting the average number of monthly occurrence days for the climatic characteristics shown in Table 10 from the total annual working days. This calculation should be based on data collected over a period of at least 5 to 10 years.

Weather Condition (5~10 Years Average Day)		Sum	Jan	Feb	Mar	~	Sep	Oct	Nov	Dec
Temperature	Over 30℃									
	Below 0℃									
	Below -5℃									
	Below -15℃									
Wind Velocity	Over 10 m/s									
	Over 15 m/s									
	Over 24 m/s									
	Sandstorm									
Rainfall	Over 5 mm									
Relative Humidity	Over 85%									
Snowfall	Over 1 cm									

Table 10 Climate Impact on Production Program

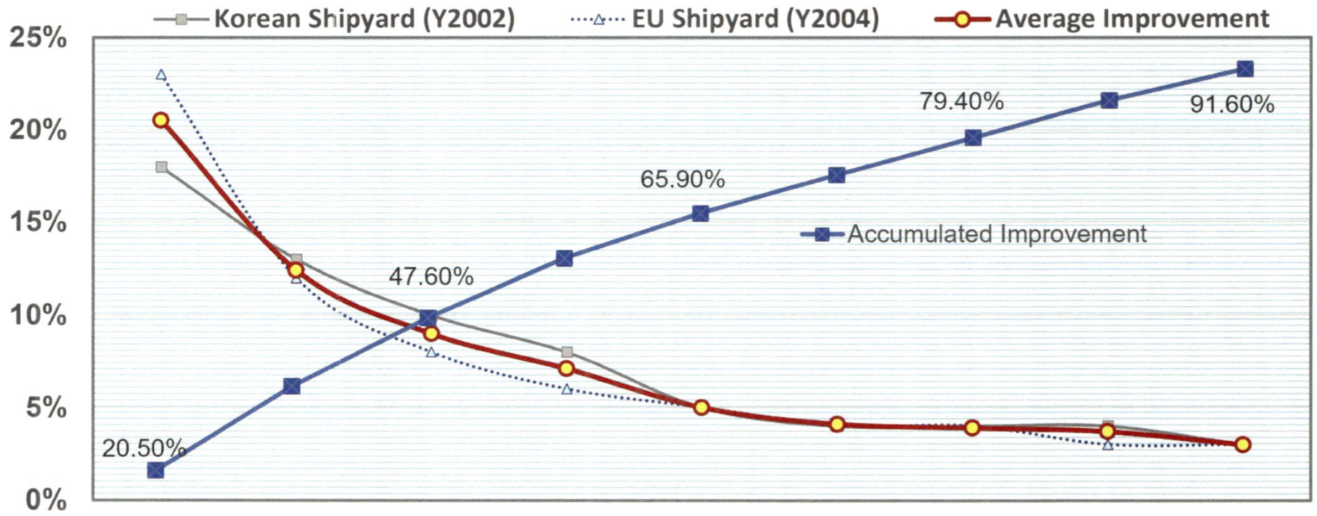

Shipyard	1st Year	2nd Year	3rd Year	4th Year	5th Year	6th Year	7th Year	8th Year	9th Year
Korean Shipyard (Y2002)	18%	13%	10%	8%	5%	4%	4%	4%	3%
EU Shipyard (Y2004)	23%	12%	8%	6%	5%	4%	4%	3%	3%
Average Improvement	**20.5%**	**12.4%**	**9.0%**	**7.1%**	**5.0%**	**4.1%**	**3.9%**	**3.7%**	**3.0%**
Accumulated Improvement	20.5%	35.4%	47.6%	58.0%	65.9%	72.8%	79.4%	86.0%	91.6%

Figure 12 Typical Productivity Learning Curves

2.3 Production Program

A shipyard's mid- and long-term production program typically requires a schematic representation of the production schedule for each project over a period of 3 to 5 years in order to make it easily recognizable. As depicted in Figure 13, a new shipyard operation plan needs to be developed for a period of approximately 10 years, starting from the time of shipyard construction. The purpose of this plan is to provide raw data that can be used to assess the return on investment and corporate value in the future. When developing a production program, the target productivity should be set by referring to the productivity of the major shipbuilding countries mentioned above, and then the shipbuilding period and production man-hours for each project of the new shipyards should be determined.

In the meantime, in addition to setting production goals, the production program becomes the reference point for shipyard design and construction during this period.

Of course, each shipyard may have different standards for assigning a project ID number when developing a production program. However, Korean shipyards typically assign a project ID number consisting of 3 to 4 digits. This number is determined based on the characteristics of each ship type and the serial number of the project award, as described below. This allows for more effective communication of extensive project information among all members of the shipyard.

- 1000: Bulk Carrier
- 2000: Gas Carrier
- 3000: Ro-Ro Ship
- 4000: Container Ship
- 5000: Oil Tanker
- 6000: Offshore Product
- 7000: Special Ship (Naval Ship, Coast Guard Ship)
- 8000: Repair Ship
- 9000: Special Purpose Use

Meanwhile, since it is not easy to expand dry dock facilities to increase sales after the shipyard's productivity reaches a certain level, increasing the dry dock turnover ratio is the most important factor for productivity innovation. Figure 14 illustrates the lead time between each production process when four Crude Oil Tankers (COTs) are constructed simultaneously using a tandem and parallel construction method at a large dry dock in a Korean shipyard. This will not be discussed in detail here, but it is a level that can only be achieved if the shipyard's operating system and the infrastructure of the shipbuilding industry are well established.

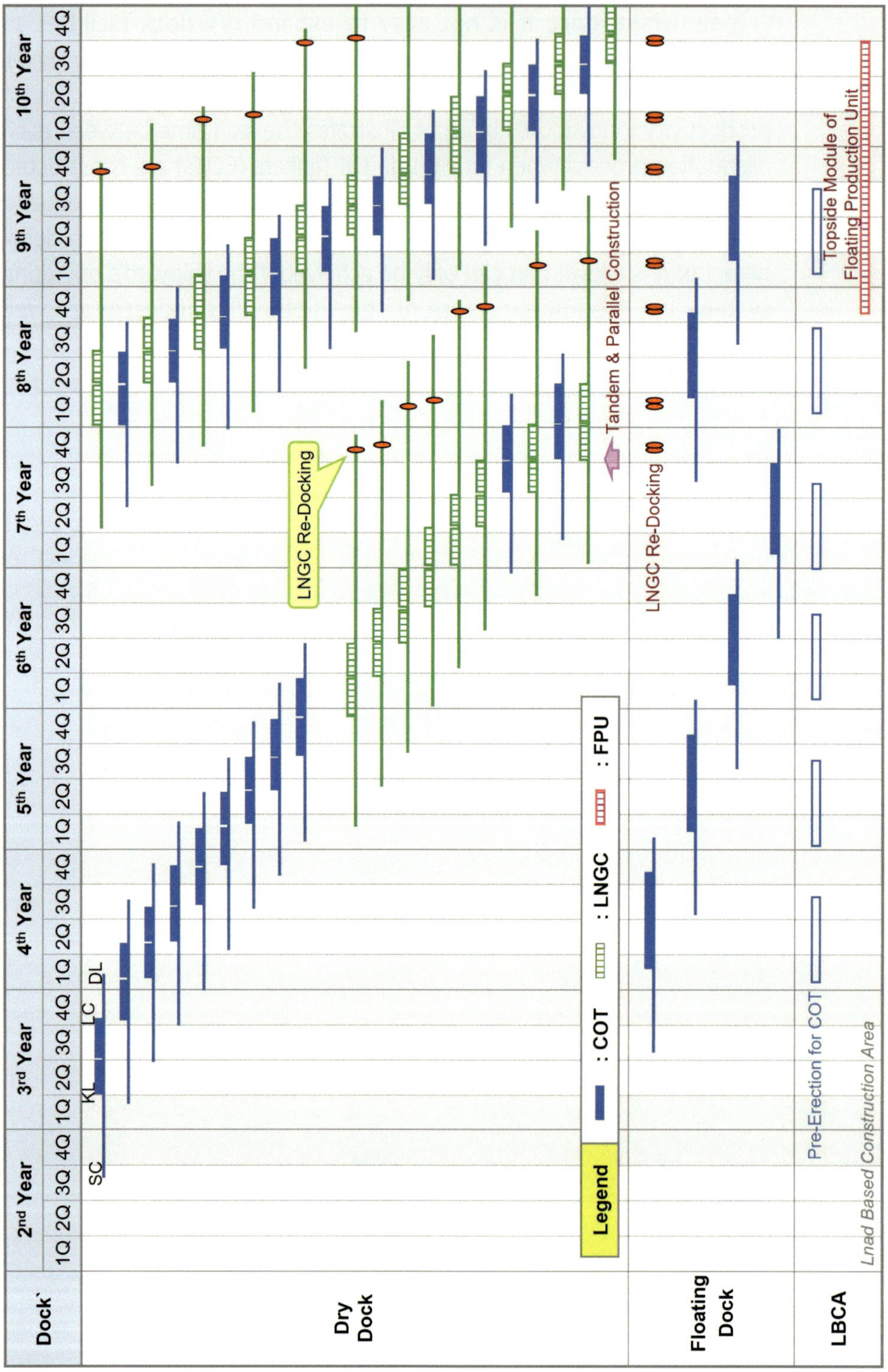

Figure 13 Typical Production Program

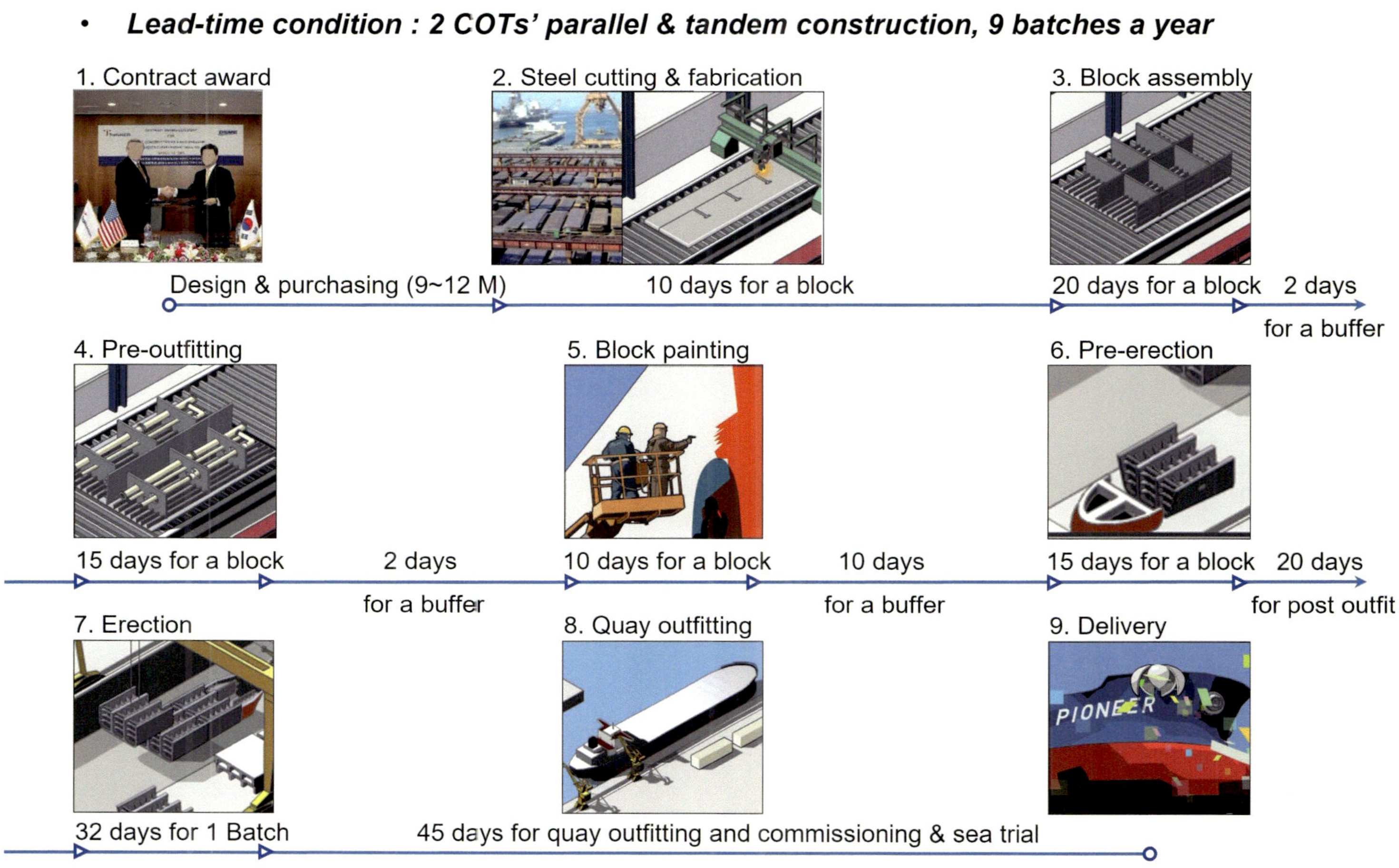

Figure 14 Typical Lead Time by Production Process

3 Capacity of Each Production Process

From now on, it is necessary to first determine the production capacity needed for each stage of the production process, starting from material unloading to ship delivery, referring to the production target outlined in Figure 11. In addition, it is necessary to establish production facilities that can meet the required production capacity.

3.1 Material Unloading

Depending on the type of materials delivered to the shipyard, they may be transported directly to the shipyard quay by sea or transported by land and/or railroad.

The composition ratio of steel material, which is a representative material used in shipbuilding, is as follows:

- Weight ratio of steel plate: approximately 90%
- Weight ratio of steel profile: approximately 10%

In the case of domestic shipyards, considering productivity, they frequently utilize oversized steel materials that exceed the standard size limit for container transportation. However, this may be limited depending on the production capacity of the steel mill that the shipyard primarily works with and the shipyard's own production facilities.

- Usage ratio of standard size (L12 x W2.34 m, with a maximum loading weight of 21 tons): Approximately 15%
- Usage ratio of non-standard size (L22 x W4.2 m ~ L12 x W2.34 m): Approximately 75%

In the case of standard sizes, transportation by land or railway is possible, but for sizes beyond that, maritime transportation is generally used. Therefore, the steel unloading quay should be designed with reference to the specifications of the ship used for such transportation.

- Bulk carrier (Deadweight 12,000 tons): L160 x W20 x T (Draft) 8.5m
- Dedicated vessel for steel transportation (Deadweight 7,400 tons):

L126 x W20 x T6m, equipped with a specialized loading and unloading cassette.

- Barge for heavy equipment transportation:
  L85 x W27 x D6.3m (for the main engine)
- Unloading Quay Facilities:
  - Quay length: Ship's LOA (Length Overall) + ship's width + α (margin)
  - Quay water depth: Ship's draft + UKC (Under Keel Clearance, 10% of draft) + α
  - Quay crane: Utilize a magnetic crane or mobile crane that can safely handle the weight based on the size of the steel being transported.

3.2 Steel Stock Yard (SSY)

The steel stock yard is a facility where plates and section profiles are stored, and steel materials intended for production are classified and labeled. The ideal capacity of this steel stock yard would match the total output of one rotation of the entire dry dock in the shipyard. However, it should include a surplus equal to the variation in the lead time for steel material procurement or the shipbuilding period. Furthermore, if steel materials are stored under bare conditions for an extended period of time, corrosion phenomena such as pitting can worsen. Hence, the optimal level of consumption is estimated to be around 3 months, which is divided into four parts based on annual steel consumption.

Therefore, for instance, in the case of a shipyard producing 200,000 tons of assembly volume per year, the design of the steel stock yard can be arranged according to the sample layout shown in Figure 15, taking into consideration the following factors.

- Total annual capacity of SSY: 220,000 tons
  (Apply an average scrap ratio of 10%.)
- Annual turnover: 4 times
- Amount of steel stock: 55,000 tons
  - Plate (90%): 49,500 tons
  - Profile (10%): 5,500 tons

- Available area ratio of SSY: 50%,
- Stacking height: approximately 1.0 m

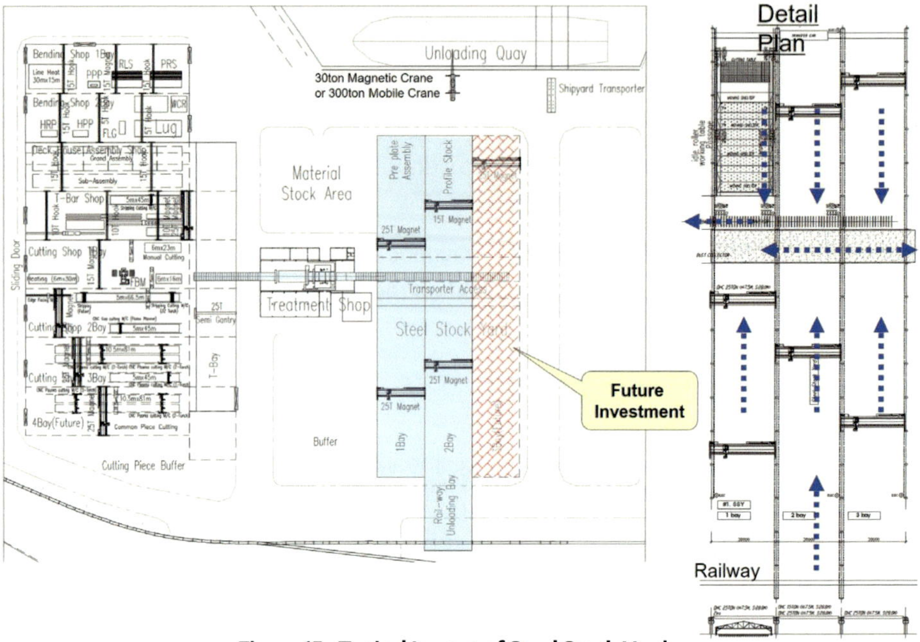

Figure 15 Typical Layout of Steel Stock Yard

- Common space between bays (transfer space in the center part of the SSY): Transporter access road + 2 rows for plate sorting + roller conveyor space (installed up to the steel treatment and cutting workshop) = 4 x (maximum width of steel plate + α)
- Bay width of SSY (two rows of standard size steel plates in the longitudinal direction): 29.6 m (approximately 30 m) = 2 x 0.45 m (for the traffic space on the upper part of the bay pedestal) + 2 x 0.5 m (from handrail to crane rail) + 27.7 m [rail span = 2 x 12 m (2 rows of standard length plates) + 3 x 0.9 m (3 rows of passageway) + 2 x 0.5 m (from crane rail to inside of pedestal)]
- Storage of section profiles: Utilizes a dedicated area of SSY that is separate by the roller conveyor.
- Plate pre-assembling area: Plate pre-assembling work is generally not performed in overseas shipyards. However, this work is an essential process for the Build Method, which necessitates a high level of accuracy control. In particular, it is a prerequisite process that must be applied for the pull-to-fit method of transverse members in the panel

block workshop. Therefore, in this work area, after butt welding the transverse members that have multiple slot holes, the slot holes can be cut at the cutting workshop later. In addition, even in the case of small internal materials with varying plate thicknesses, it is possible to cut them after plate butt welding instead of assembling them after cutting each individual member. This working area utilizes one section of the steel stock yard, which is separated by a roller conveyor, similar to the storage of section profiles.

- Required length of SSY: Approximately 3 x 180 m
  - [(Net area of SSY / 0.5) / Bay width / Number of bays] + Length of plate pre-assembling area,

  or another idea is as follows.

  - Plate storage area + Section profiles storage area + Plate pre-assembling area + Bay common space

- Steel Stock Yard Facilities:
  - Magnetic (or vacuum) crane: The safety working load of the crane hoisting is selected to have a lifting capacity of 130% or more compared to the maximum weight of the plate used, taking into account the magnetic device. However, it is reasonable to use a crane with a relatively small capacity when considering the shape of the material in the area where the section profiles are being handled. The number of operating cranes required is at least one crane for each zone separated by a conveyor, meaning that there should be two cranes per bay. If the bay length of the steel stock yard is over 200 m, it may be necessary to increase the number of cranes operated. A crane rail height of approximately 7.5 m is an appropriate level.

  - Main facilities for the plate pre-assembling area: Magnetic crane, semi-automatic cutting machine, SAW welder, and rollable workbench (22 x 40 m)

  - Roller conveyor (from SSY to T-Bay in the cutting workshop): A width of approximately 5m, capable of transporting standard-sized plates in two rows in the width direction, is appropriate.

  - Control system for SSY: With the recent development of ICT technology, this process is recognized as a representative production process that can be fully automated and unmanned in shipyard production.

Meanwhile, Korean shipyards often subcontract specialized blocks, such as deckhouse blocks, to third-party companies. During this process, not only is the storage of steel materials outsourced, but also the pre-treatment process, which is performed by logistics companies located near the steel mill. The pre-treated steel materials are then supplied directly to the outsourcing companies for block assembly works.

3.3 Steel Treatment Workshop

As described above, the steel pre-treatment process follows a continuous flow in the following order: Steel Stock Yard → Heating → Blasting → Spraying → Drying → T-Bay of Cutting Workshop. If the shipyard's annual steel material processing capacity does not exceed 250,000 tons, it is reasonable to have a facility that can handle both plate and section profiles, as shown below. Figure 16 shows the sample layout of a workshop that includes a steel treatment process.

- Dual-use for plates and profiles
  - Plate: t = 6 ~ 150 mm, W = 900 ~ 4,500 mm, L = 23 m
  - Profile: H-Beam = H600 x W300 mm, Angle = Min. 50 x 50 mm ~ 450 x 200 mm
- Steel Treatment Workshop Facilities: Preheater, Shot Blaster, Dust Collector, Paint Booth, Drying Booth, Roller Conveyor, Auto Marking System, and a 2.8-ton OHC for Maintenance
- Workshop size: L60 x W40 x H12.5 m

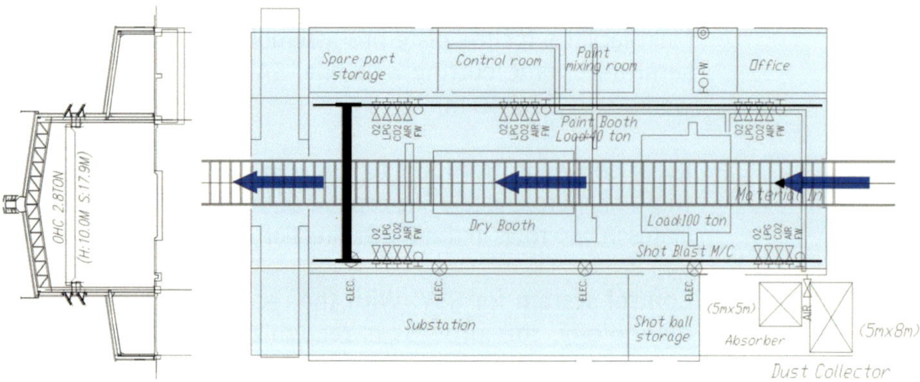

Figure 16 Typical Layout of Steel Treatment Workshop

3.4 Cutting Workshop

Except for manual cutting for section profiles, shipyards generally use three types of CNC (Computer Numerical Control) cutting machines. Although there may be variations in the cutting volume ratio depending on the type of ship, Table 11 displays the average ratio of cutting volume by cutting type at each workshop, as well as the unit processing capacity by cutting machine. Based on this empirical data, it is possible to calculate the number of cutting machines required to process 220,000 tons of steel per year, which is assumed to be the production goal. It is necessary to consider some flexibility in operations by adjusting the equipment operation rate to account for any excess or deficiency of required equipment.

<table>
<tr><th>Type of Machine</th><th>Use</th><th>Cutting Volume Ratio</th><th>Unit Capacity per Year</th><th>Required Cutting Volume</th><th>Required Quantity</th></tr>
<tr><td>Plasma Cutting Machine</td><td rowspan="4">For Cutting Workshop</td><td rowspan="2">52%</td><td rowspan="2">21,000</td><td rowspan="2">114,400</td><td rowspan="2">5.4</td></tr>
<tr><td>Flame Planner Cutting Machine</td></tr>
<tr><td>Stripping Cutting Machine</td><td>8%</td><td>9,450</td><td>17,600</td><td>1.9</td></tr>
<tr><td>Profile Cutting (Manual Cutting)</td><td>10%</td><td>27,300</td><td>22,000</td><td>-</td></tr>
<tr><td>Stripping Cutting Machine</td><td>For T-Bar Workshop</td><td>6%</td><td>55,000</td><td>13,200</td><td>0.2</td></tr>
<tr><td>Flame Planner Cutting Machine</td><td>For Panel Block Workshop</td><td>24%</td><td>42,000</td><td>52,800</td><td>1.3</td></tr>
</table>

* *Note: The required cutting volumes and quantities are based on a processing weight of 220,000 tons*

Table 11 Cutting Volume Ratio by Workshop and Cutting Capacity

Transfer Bay

As depicted in Figure 17, the T-Bay is an area where steel, transported via a roller conveyor from the steel treatment workshop, is sorted before being transported to the subsequent cutting process. A semi-gantry will be installed in this area to facilitate movement to the entrance of each cutting bay assigned by the cutting type. It must be installed below the overhead crane to allow movement across the cutting bays. The roller conveyor installed in this T-Bay area tilts in one direction to facilitate the lowering of the steel plate onto the T-Bay floor, thereby reducing the workload on the crane In addition, it is desirable to arrange the T-Bay area to provide the shortest logistics flow between the T-Bar workshop, which produces T-Bars by simply cutting steel plates into strips, and the deckhouse assembly workshop, which produces blocks using just steel plates, T- Bars, and angles.

In the meantime, these workshops have the lowest height among all the workshops in the shipyard.

- Transfer Bay Facilities:
    - Semi-gantry magnetic crane: It must have the same capacity as the maximum capacity of the crane in the steel stock yard, and it must also be equipped with a hook device to transfer the section profiles and flat bars in a bundled state.
    - Auto transfer equipment: Chain transfer conveyor combined with roller conveyor.

Cutting and Bending for Section Profiles

In this bay, the cutting process of section profiles such as angles, bulb plates, T-Bars, etc., as well as the bending process of section profiles to be attached to curved shell plates. Section profile cutting is generally performed manually, rather than using automated cutting methods. There are two methods for bending work: one method utilizes hydraulic machines only, while the other involves additional line heating to achieve the desired curvature. One row of this bay serves for a simple cutting process and a temporary buffer area for palletizing. The other row is arranged to enable line production in the following order:

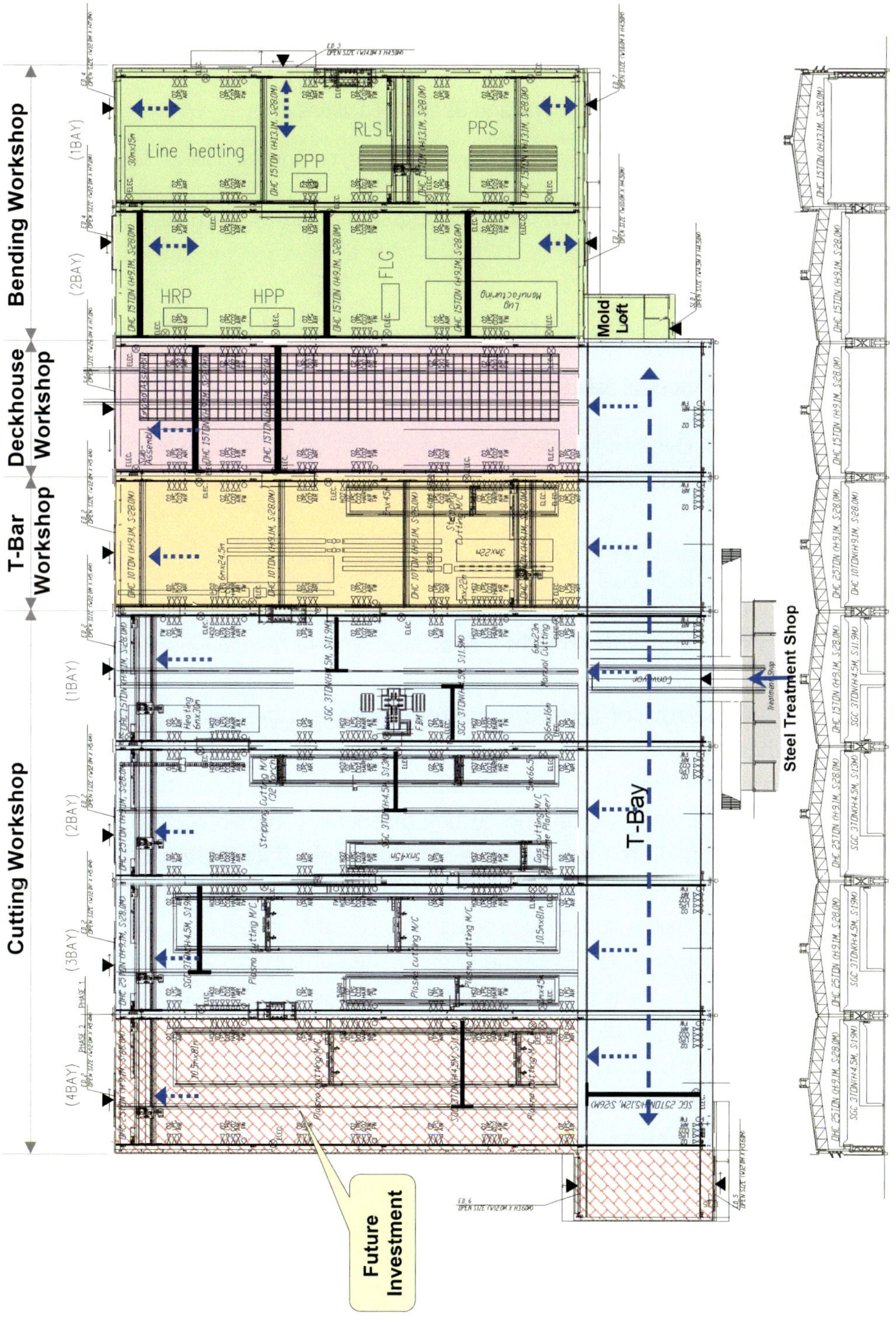

Figure 17 Typical Layout of Cutting Workshop (including T-Bar, Deckhouse Assembly, and Bending Workshop)

"Section profile cutting → Primary bending using FBM (Frame Bending Machine) → Final curvature fabrication by line heating".

- Section Profile Cutting and Bending Facilities:
  - FBM: Max. bending force 700 tons, compressed air of 7 bar
  - Workbench for line heating: ϕ50 mm perforated holes with a 600 mm interval
  - Manual cutting table: 2 sets
  - Special pallet for profile handling
  - Overhead crane: The same capacity as SSY's crane for the profile stock area
  - Semi-gantry magnetic crane: Small capacity crane for handling individual section profiles

Plate Cutting with Straight Lines

In one row of this cutting area, there is a designated work area where a variety of components, such as stiffeners, carlings, and face plates, are produced. These components are made by cutting plates at close intervals using a strip cutting machine. In the adjacent row, there is another work area where even extremely thick plates, measuring 100mm or more, can be cut using a flame planner cutting machine. The final step in the stripping process is to use the edge-facing machine on the conveyor to fabricate the edge of the cut surface. This ensures that there are no defects in the quality of the painting. Additionally, the size of the cutting table should be sufficient to ensure uninterrupted cutting operations without the need to transfer cut members.

- Plate Cutting Facilities (for straight lines):
  - Strip cutting machine: 32 cutting torches
  - Flame planner cutting machine: 4 cutting torches
  - Cutting table: 2 sets

- Edge-facing machine for flat bars
- Overhead crane: The same capacity as T-Bay's overhead crane
- Semi-gantry magnetic crane: Small capacity crane for handling individual stripped members

Plate Cutting with Various Shapes

This cutting bay is a designated work area where plates are rapidly cut into various shapes using a plasma cutting machine. To maximize cutting efficiency, it is advisable to arrange the cutting tables in a 2x3 matrix in one row to cut the largest size plate, and arrange the cutting tables in a 1x2 matrix in the opposite row. Additionally, it is necessary to equip this cutting machine with a device that automatically displays the unique identification number of the cut piece.

- Plate Cutting Facilities (for various shapes):
  - A-type plasma cutting machines: 2 cutting torches
  - B-type plasma cutting machines: 1 cutting torch
  - Cutting table: 2 sets
  - Overhead crane: The same capacity as T-Bay's overhead crane
  - Semi-gantry magnetic crane: Small-capacity crane for handling individual cut pieces

For reference, the plasma underwater cutting machine has the advantage of minimizing plate deformation caused by cutting heat. However, except for relatively thin plate products such as passenger ships, naval ships, and car decks of Ro-Ro ships, it is generally not widely used for constructing other types of merchant ships due to productivity concerns.

On the other hand, a buffer area equivalent to 20 to 25% of the total area of the cutting workshops is generally required around this workshop for sorting and palletizing cut pieces for transfer to the next process.

3.5 T-Bar Manufacturing Workshop

The production process of the T-Bar manufacturing workshop has been explained in detail in Section 1.5 of this Part II. However, the arrangement of equipment in this workshop and the production flow are described in more detail in Figure 18 below. In Korea, there is a specialized manufacturing company that directly supplies T-Bars to outsourcing companies involved in block assembly.

- T-Bar Manufacturing Workshop Facilities:
  - ① Strip cutting machines: 20 Cutting torches
  - ② Face plate standby area
  - ③ Web plate standby area
  - ④ T-Bar fit-up machine: Face setting stroke of 220 mm, web setting stroke of 50 mm
  - ⑤ Tag welding machine: Max. feeding speed of 25 m/min
  - ⑥ Welding machine: Max. feeding speed of 25 m/min
  - ⑦ Straightening machine: Straightening of bent and deformed T-Bar
  - ⑧ Edge facing machine: Max. feeding speed of 20 m/min
  - ⑨ Blasting and painting machine: Max. feeding speed of 10~15 m/min
  - Overhead crane for plate handling: The same capacity as the T-Bay's overhead crane
  - Overhead crane for T-Bar handling: 3 small-capacity overhead cranes

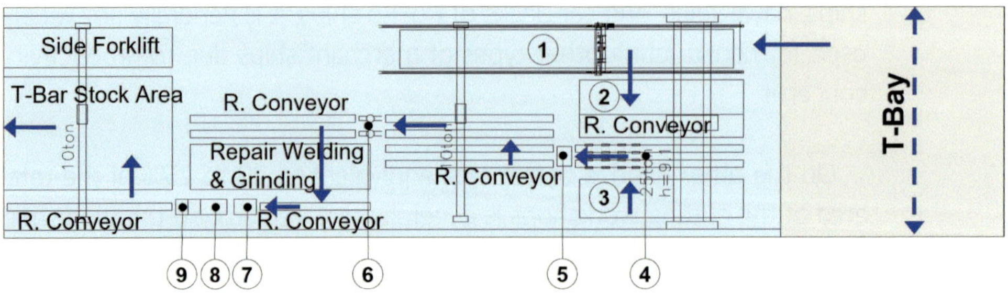

Figure 18 Production Flow of T-Bar Manufacturing Workshop

3.6 Bending Workshop

In most cases, the cut members are moved to the next process, which is the minor assembly process, immediately after cutting. However, depending on the type of member, some of them may require a bending process beforehand. Various cold bending equipment and line heating methods are used in this process, depending on the size or type of the cut member. The typical layout of the bending workshop and the workflow are shown in Figure 17, which was described previously.

Large-Sized Plate Bending

The workshop height should be sufficient for processing large-sized components, such as curved shell plates, corrugated bulkheads, and hopper structures in cargo holds. A workbench for final line heating, after primary bending with cold bending equipment, should be positioned on the exit side of the bay. Additionally, a separate mold loft should be arranged nearby, equipped with a CNC wood cutting machine, to produce wooden templates for quality control of line heating work.

As a reference, the workload for this bending process accounts for approximately 3.0% to 3.1% of the total number of structural members, calculated based on the annual steel material throughput.

- Large-Sized Plate Bending Facilities:
  - 1500-ton Plate Rolling Press Machine: Max. width 12 m, up to 43 mm thickness
  - 1500-ton Knuckle Press Machine: Max. width 13 m, working height 850 mm
  - 1000-ton Pin Press Machine: Bed size of 6,000 x 2,000 mm
  - Workbench for line heating: ϕ50 mm perforated holes with a 600mm interval
  - CNC Wood Cutting Machine:
    Applicable size of L2,600 x W2,500 mm

- Overhead Crane: 2 Hook cranes for handling a 12 m long steel plate, and 1 magnetic crane

Small-Sized Member Bending

Among the small components commonly needed in the process of block assembly and erection, there is a need for a work area to cold bend relatively small flange brackets, as well as curved and/or knuckled face plates. Additionally, a workspace is needed to manufacture collar plates, lifting lugs, and other items required in large quantities using scrap materials. It is advisable to have a work area for these small components located next to the cutting workshop. This way, they can be fabricated using scrap and leftover plates that are generated during the cutting process.

- Small-Sized Members' Bending Facilities:
  - 640-ton Press Brake Machine: For fabricating flange brackets and swage bulkheads, with an allowable working length of 5,100 mm
  - 400-ton Horizontal Roll Press: For fabricating curved face plates, with an allowable working width of 1,100 mm
  - 300-ton Horizontal Pin Press: For fabricating knuckled face plates, with an allowable working width of 1,100 mm
  - Manual Cutting Machine and Workbench
  - Overhead Crane: 1 Overhead crane for flanging work, and 2 small capacity overhead cranes for other tasks

3.7 Analysis of Block Assembling Capacity

The hull block is divided into three categories based on the type of block: panel block, curved block, and deckhouse block with funnel casing. The composition ratio of these blocks varies significantly depending on the type of ship being built. In addition, the size of a unit block may vary depending on the facilities of the assembly workshop and how the block is divided according to the block division drawing. However, Table 12 below serves as a reference index, showing the weight ratio by block type based on the performance of large shipyards in Korea.

Categorization		LNGC	Suezmax COT	Aframax COT
Panel Block	Weight Ratio	54.2%	68.4%	69.1%
	Avg. Weight	165 tons	161 tons	134 tons
Curved Block	Weight Ratio	40.4%	28.7%	27.8%
	Avg. Weight	102 tons	99 tons	99 tons
Deckhouse and Funnel Casing	Weight Ratio	5.4%	2.9%	3.1%
	Avg. Weight	52 tons	47 tons	41 tons

Table 12 Weight Ratio by Block and Ship Type

Additionally, the number of small components with simple shapes to be manufactured before the minor assembly process is equivalent to approximately 2% of the total assembly weight. Some shipyards even have separate workshops to effectively assemble these components. However, in other cases, workspace is allocated for the manufacture of these components within the three types of block assembly workshops, with a priority on minimizing logistics movements to the next process. Here, the latter case is taken as a reference, and the relevant assembly workshops are subsequently described.

Even with the previously assumed annual assembly production capacity of 200,000 tons, it should be noted that the production capacity required for each assembly workshop may vary depending on how the product mix of the shipyard is configured. Table 13 provides a sample calculation of the production capacity required for each assembly workshop, assuming the annual construction of 3 LNGCs and 6 Suezmax COTs, based on the reference index mentioned above. The panel block workshop, which produces the largest amount of assembly work, is based on a line production by tact time with two shifts per day. For reference, in large Korean shipyards, the tact time is approximately 3.5 hours.

Categorization	PBS	CBS	DBS
Block Weight Ratio	63.2%	33.0%	3.8%
Annual Production Volume	126,400 tons	66,000 tons	7,600 tons
Average Weight of a Block	163 tons	101 tons	50 tons
Annual Working Day	240 days		
Number of Blocks per Day	3.2 ea	2.7 ea	0.6 ea

* *Note: PBS - Panel Block Workshop, CBS - Curved Block Workshop, DBS - Deckhouse Block Workshop*

Table 13 Required Production Capacity by Assembly Workshop

On the other hand, while it is highly desirable to proceed to the next pre-outfitting process immediately after completing block assembly, temporary waiting may be unavoidable due to several factors, including progress delays in subsequent processes such as pre-outfitting, painting, pre-erection, and erection processes. When the annual dry dock turnover rate is increased to expand sales and improve overall productivity in the shipyard, there may be a risk of serious overload in the block buffer area. Therefore, if the buffer area for the assembled block, excluding the block in progress, is not properly allocated around the grand assembly workshop, there may be a possibility that the overall assembly process could be halted. Therefore, it is recommended to allocate a temporary buffer area around each block assembly workshop, which is able to accommodate a number of blocks equivalent to 1% to 2% of the total number of assembly blocks produced per year.

3.8 Panel Block Workshop (PBS)

In the previous overall description of the shipyard production process, the role of the panel block workshop for each working bay was briefly explained. However, Table 14 and Figure 19 provide a sample layout of this workshop, based on the premise that it has the panel block assembly capacity defined in Table 13 above.

Based on empirical data, it is approximately 0.06~0.07 m^2 of work-space is required for the production of 1 ton of minor and sub-assembly work needed for panel block assembly from Bay 4 to Bay 6.

After completing the final welding of the panel block and the installation of hull structuralized outfittings in Bay 7, according to the tact time, the block is transferred outside of Bay 7 along the skidway using the transport carriage located at the bottom of the block.

Bay ID	Key Role	Bay Size
Bay 1	Edge cutting of primary plate	L107 x W28 x H12m
Bay 2	One-side welding of primary plate	L32 x W32 x H12m
Bay 3	Turning over the primary plates, welding the back-side, and auto-marking in the longi-position	L32 x W32 x H38.3m
Bay 4	Supplying longi-members like T-Bars, bulb plates or angles and welding longi-members, components, and minor assemblies on the primary plate	L152 x W32 x H13.6m
Bay 5	Supplying trans-members, then positioning them on longi-members using the pull-to-fit method, and partially welding for the trans-members	L152 x W45 x H26.8m
Bay 6	Reinsertion of a single hull block and turning over this block, then positioning it using the pull-to-fit method to complete the double hull block, and partial welding of single and/or double hull block.	L152 x W45 x H38.3m
Bay 7	Full welding for single and/or double hull block and installing the hull structuralized outfittings	L128 x W32 x H26.8m

* *Note: Height of building means the clear height beneath the roof structure.*

Table 14 Size and Key Function by Each PBS Bay

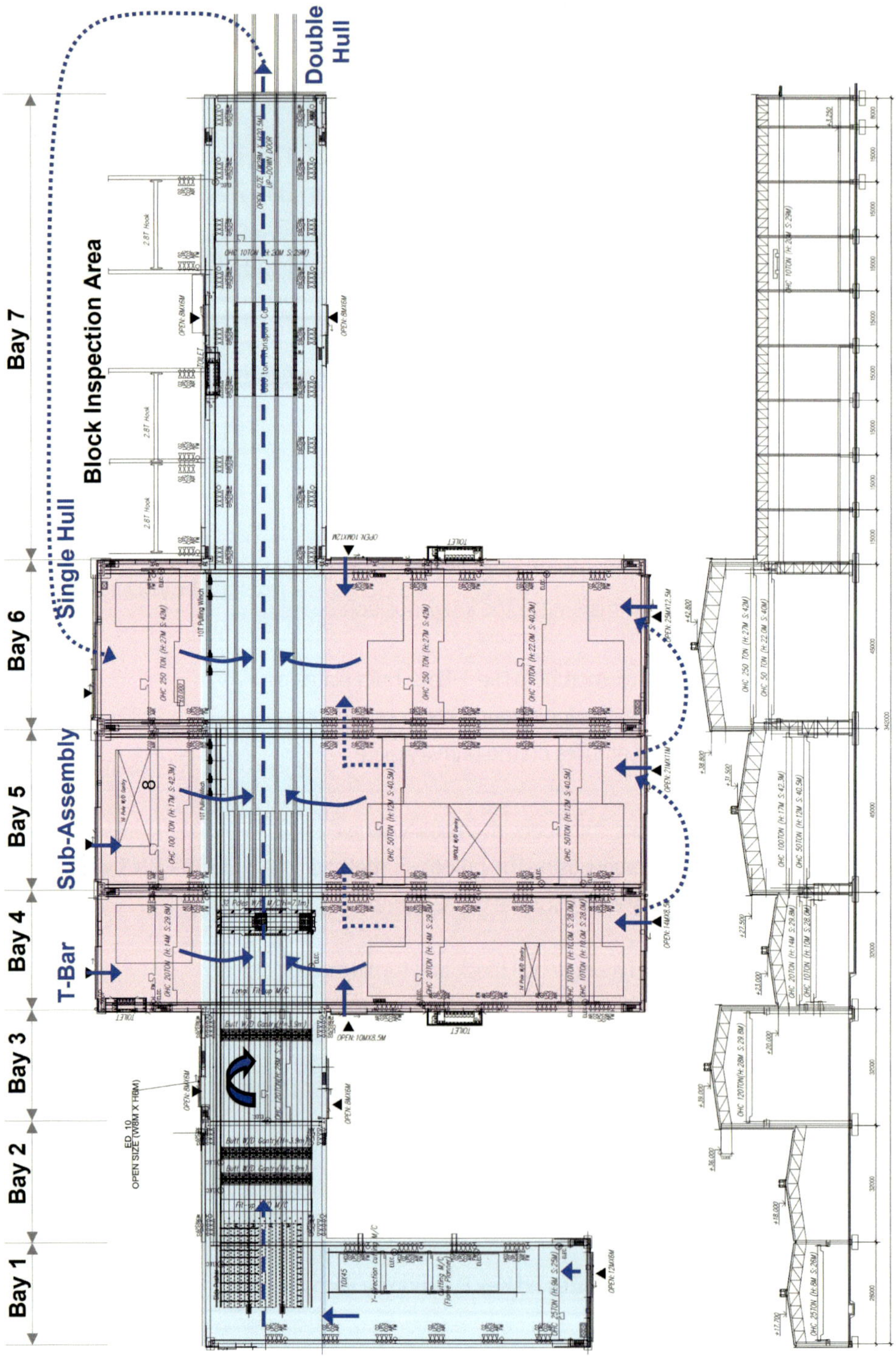

Figure 19 Typical Layout of Panel Block Workshop

In order to assemble a double hull block, the initially assembled single hull block must be reinserted into Bay 6. The completed single hull or double hull block is then transported to the block inspection area, which is located near the panel block workshop. Afterwards, finishing work, such as grinding, is completed, and inspection work is subsequently performed by the shipowner and classification society. By doing this, there will be no restrictions on the tack time of the line production system in the panel assembly workshop. Therefore, the number of block inspection areas must be calculated, taking into account the daily block assembly throughput and the lead time of the finishing process in the block inspection area.

- PBS Bay 1 Facilities:
  - Flame Planer Cutting Machine: 4 torches, 1 set
  - Y-Direction (vertical) Cutting Machine: 1 torch, 1 set
  - Chain Conveyor: from Bay 1 to the middle of Bay 5, equipped with idle rollers
  - Hydraulic Push System
  - Overhead Magnetic Crane: 1 set for handling the weight of primary plates

- PBS Bay 2 Facilities:
  - Plate Fit-up Machine: 1 set
  - SAW Butt Welding Gantry: 2 sets for one-side welding

- PBS Bay 3 Facilities:
  - SAW Butt Welding Gantry: 1 set for back-side welding
  - Overhead Crane: 1 set for turning over welded primary plates, equipped with a block loader

- PBS Bay 4 Facilities:
  - Fillet Welding Gantry: 32 poles, 1 set for welding longitudinal members, operated from Bay 4 to Bay 5
  - Fillet Welding Carriage: 16 poles, 1 set for welding minor and sub-assemblies
  - Fit-up Machine: 2 sets for the alignment of longitudinal members
  - Overhead Crane: 2 sets for component and minor assembly, and 2 sets for sub-assembly

- PBS Bay 5 Facilities:
  - Fillet Welding Carriage: 16 poles, 2 sets for welding sub-assemblies
  - Pulling Winch: 4 sets
  - Overhead Hook Crane: 2 sets
  - Overhead Turning Crane: 1 set, equipped with a block loader

- PBS Bay 6 Facilities: Application of welding robots for confined space work
  - Pulling Winch: 4 sets
  - Overhead Crane: 1 set
  - Overhead Turning Crane: 2 sets, equipped with block loaders

- PBS Bay 7 Facilities:
  - Transport Carriage: 4 sets, skidway from Bay 5 to outside of Bay 7
  - Overhead Hook Crane: 1 set for handling pre-outfit materials

- Block Inspection Area Facilities:
  - Overhead Hook Crane: 1 set for each block inspection area

3.9 Curved Block Workshop (CBS)

For an overview of the production process of the curved block assembly, refer to Section 1.5 of Part II. Additionally, the factors that determine the workshop layout, as shown in Figure 20, will be described here.

The size of the curved block assembly workshop is primarily determined by the number of workbenches needed for the grand assembly works at Bay 3. To calculate the minimum number of workbenches required, it is necessary to consider the lead time and daily throughput of the curved block workshop, according to the analysis of the block assembling capacity calculated in Table 13.

For reference, the lead time for the grand assembly of the curved block in Korean shipyards is approximately 3.4 days. The size of Bay 3 should be determined by carefully comparing and evaluating the local productivity before making any necessary adjustments. In other words, if this workshop has to assemble 2.7 curved grand assembly blocks daily, based on the previous assumption and taking into account the local productivity target being 2/3 of that in Korea, the estimated lead time for this workshop will be approximately 5.1 days. Consequently, approximately 14 workbenches are needed in Bay 3. Since there are variations in the sizes of large, medium, and small blocks, it is necessary to determine the dimensions of Bay 3 by appropriately combining the sizes of each workbench. Additionally, a vacant space the size of one workbench must be allocated for the transporter's accessibility at the entrance and exit of Bay 3.

Component, minor, and sub-assembly works performed in Bay 1 and 2 require a larger area compared to the case of a panel block. According to empirical data, it is estimated to be approximately 0.12~0.33 m^2 to fabricate 1 ton. Unlike the case of panel block assembly, such a significant deviation in the required area means that it takes a considerable amount of time to improve the productivity of the curved block assembly workshop, as well as the variation in shape of the members. Therefore, when establishing a new shipyard, it is desirable to prioritize the higher value among these reference values.

In the case of a shipyard that directly performs the assembly of large outfittings, such as the rudder and the assembly of the funnel casing, without outsourcing to a specialized manufacturer outside the shipyard, these assembly works are usually carried out in Bay 2 of the curved block workshop.

As is the case with panel block assembly, the curved block that has been assembled is then transferred to the block inspection area. Here, it undergoes finishing work, including grinding, before being inspected by shipowners and classification societies. Therefore, considering the lead time required for the finishing work, an appropriate number of block inspection areas is placed adjacent to the curved block workshop.

Meanwhile, due to the rapid development of ICT, there is the possibility of automatically adjusting the height of the pin jig installed on the Bay 2 and Bay 3 workbench. If this type of pin jig is used in conjunction with block assembly drawings, productivity can be greatly improved.

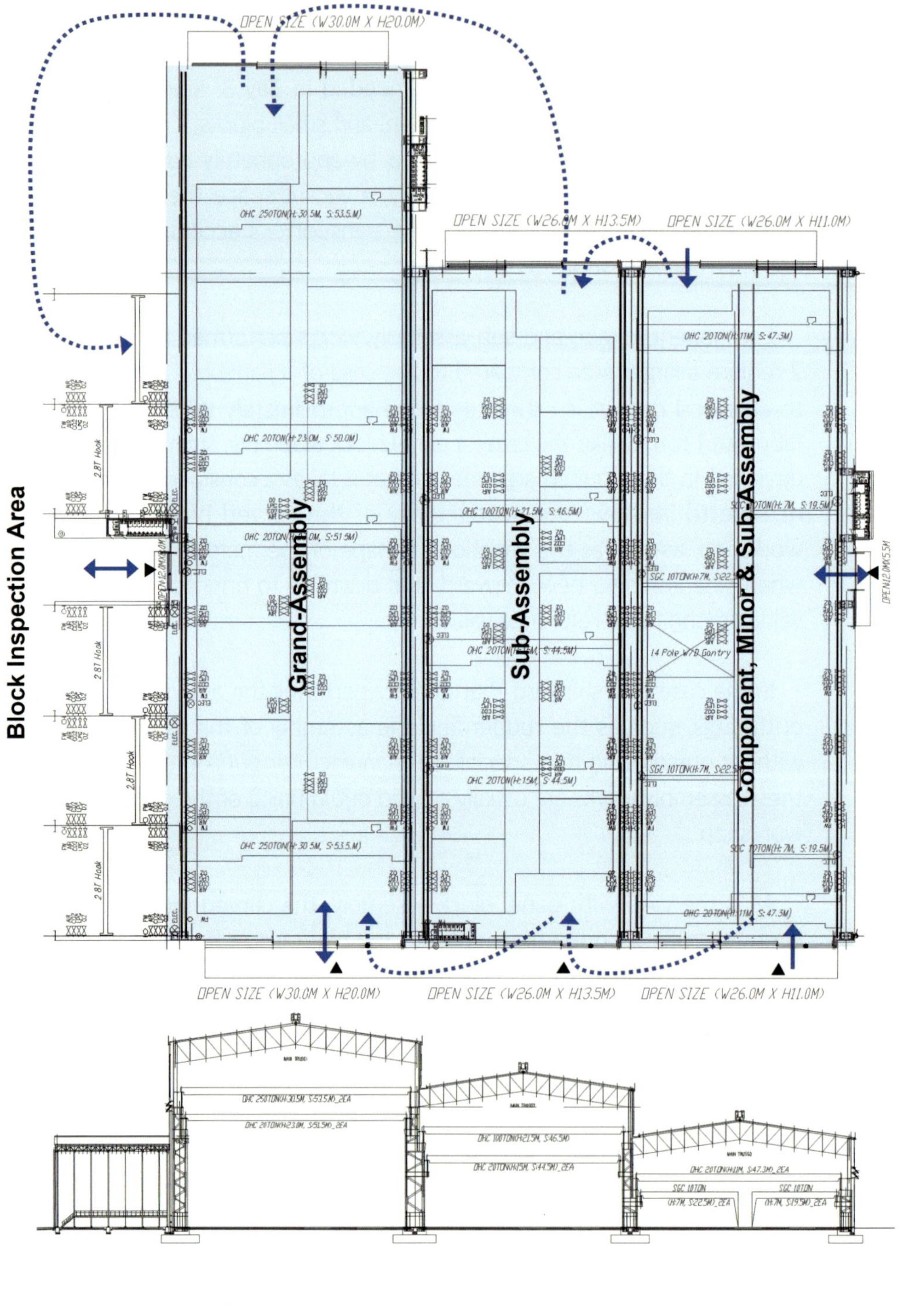

Figure 20 Typical Layout of Curved Block Workshop

- CBS Bay 1 Facilities:
  - Fillet Welding Carriage: 14 poles, 2 sets
  - Workbench with Pin Jig: Auto and manually adjustable type
  - Semi-Gantry Hook Crane: 4 sets
  - Overhead Hook Crane: 2 sets
- CBS Bay 2 Facilities:
  - Workbench with Pin Jig: 2 sets for rudder assembly and funnel casing grand assembly
  - Overhead Hook Crane: 2 sets
  - Overhead Turning Crane: 1 set, equipped with a block loader
- CBS Bay 3 Facilities:
  - Workbench with pin jig: Empty spaces for the entry of the transporter should be positioned on both sides of this bay.
  - Overhead Hook Crane: 2 sets
  - Overhead Turning Crane: 2 sets, equipped with block loaders
- Block Inspection Area Facilities: Same as the PBS case
  - Overhead Hook Crane: 1 set for each block inspection area

3.10 Deckhouse Block Workshop (DBS)

As mentioned earlier, if there are few outsourcing companies available for thin plate assembly near the shipyard, it becomes essential to have the necessary facilities within the shipyard during its initial stages of operations. As depicted in Figure 17, the deckhouse assembly workshop is relatively small in size. Therefore, it would be more appropriate, from an architectural design perspective, to locate it alongside other small-sized workshops, such as a cutting workshop. In addition, if outsourcing becomes an option in the future, this workshop can be repurposed based on the shipyard's facility expansion plan.

- Deckhouse Block Workshop Facilities:
  - Workbench with Pin Jig: Calculate the number of workbenches using the same concept as in the case of CBS

- Overhead Hook Crane: 2 sets
- Transport Carriage: 2 sets

3.11 Pre-Outfitting Workshop (POS)

To understand the pre-outfitting process, refer to the details provided in Section 1.5 of Part II. This section deals with the design of the pre-outfitting workshop.

The workload of pre-outfitting, which refers to the tasks completed before painting, varies depending on the type of block, leading to significant variations in lead time. Although pre-outfitting works are necessary for all hull blocks, they are partially performed during block assembly and/or are also completed in the pre-outfitting workshop after the initial pre-erection. Therefore, the average weight of the hull blocks brought into the pre-outfitting workshop differs from the average value in the previous analysis of block assembly capacity. Table 15 illustrates the procedure for determining the necessary workbench in the pre-outfitting workshop, taking into account the lead time of pre-outfitting tasks for each block type that is brought into the pre-outfitting workshop at major shipyards in Korea.

The productivity index was not separately applied in this case because pre-outfitting work is a relatively simple equipment and outfitting installation activity. If the management technique of pre-outfitting work is well established, it is possible to overcome productivity differences in a short period of time.

Categorization	Lead Time	Avg. Block Weight	Work Volume Ratio	Required Quantity of Workbenches
Engine and Machinery Spaces Block	9.6 days	177 tons	38%	200,000t x 38% / 177t / 240d x 9.6d = 17.2 ea
Other Hull Block	2.4 days	186 tons	62%	200,000t x 62% / 186t / 240d x 2.4d = 6.7 ea
Offshore Block	11 days	114 tons	-	

Table 15 Pre-Outfitting Lead Time by Block Type and Average Block Weight

These pre-outfitting workshops are typically located next to block assembly workshops, and the amount of pre-outfitting work required varies depending on the type of block. Therefore, taking into account the workload variations, Figure 21 shows that the bay length of a pre-outfitting workshop should be sized to accommodate just two to four workbenches. Additionally, it is necessary to allocate a block buffer area equivalent to 0.5 to 1.0% of the total number of assembly blocks per year in order to account for workload fluctuation. For blocks with a small work volume, pre-outfitting works are sometimes performed in these block buffer areas using a tower crane or a mobile crane.

As mentioned earlier, when it comes to unitized or modularized outfittings, it is common practice in Korea to outsource them to companies that specialize in this field in order to avoid bottlenecks in the overall production process flow. If the local outsourcing infrastructure is not well-developed, it is necessary to prepare a separate workspace in the shipyard, taking into account the lead time for each type.

Meanwhile, except for the procurement of hull fittings such as doors, hatches, ladders, etc., which are generally professionally manufactured outside of shipyards, some outfitting manufacturing workshops that need to be manufactured at shipyards will be described separately in the later section.

- POS Facilities:
  - Overhead Hook Crane: For the installation of equipment and small outfittings above each workbench
  - Pedestal Drilling Machine: 1 set for each POS Bay

Figure 21 Typical Layout of Pre-Outfitting Workshop

3.12 Blasting and Painting Workshop

The number of workshops required to perform the block blasting and painting process is calculated by considering the quantity of panel blocks and curved blocks produced per year, as well as the lead time for the blasting and painting process. However, in the case of accommodation blocks, power tool cleaning is generally permitted. Therefore, it is not separately considered when calculating the required quantity for the blasting and painting workshop. Table 16 presents an example of calculating the necessary working space, taking into account the productivity of the process in Korean shipyards.

During the winter season in Korea, a fan heater is sometimes used to speed up the drying process of the painted blocks in the block buffer area. However, in climates that are much harsher than this, the influence of weather may need to be considered separately, as shown in Table 16. Alternatively, the painting workshop may need additional facilities to reduce drying time. In some shipyards, the entire blasting and painting process may be equipped with a line production system.

Categorization	Indoor Bay	Outdoor Buffer	Required Quantity of Bay and Buffer
Lead time for Blasting	0.5 day	0.33 day	Blasting Bay = 4.4 (≈ 4 ea) = Daily Block Assembly (5.9 ea) x 0.5 day / Productivity Index (0.67) Blasting Buffer = 3.4 (≈ 3 ea) = 5.9 ea x 0.33 day / 0.67 / Weather Impact (0.85)
Lead time for Painting	2 days	2.5 days	Painting Bay = 17.6 (≈ 18 ea) = 5.9 ea x 2 days / 0.67 Painting Buffer = 25.9 (≈ 26 ea) = 5.9 ea x 2.5 days / 0.67 / 0.85

Table 16 Lead Time for Blasting and Painting

Since repetitive painting work is typically necessary 2 to 4 times in order to achieve the desired film thickness specified for the painting, it is recommended that the blasting bay and the painting bay be in a quantity ratio of 1:4. Meanwhile, as a general reference value, unlike the calculation method above, the buffer area for the blasting and painting process typically accounts for approximately 1.5% to 2.0% of the total number of blocks assembled per year. Figure 22 shows, as an example, that a significant amount of buffer space is required around the blasting and painting workshop.

In order to effectively perform blasting and painting work without the need for scaffolding installation, it is necessary to determine the size of each bay, taking into account the operating space required for the cherry picker in the workshop. Additionally, in the case of LNGC cargo hold blocks, it is desirable to pre-erect two individual flat blocks using only the transporter and to have a large bay so that blasting and painting can be carried out sequentially. The control room for various mechanical devices installed in the blasting and painting workshop is typically located in the area between the blasting workshop and the painting workshop.

- Blasting Workshop Facilities:
  - Dump Hopper, Screw Conveyor, Bucket Elevator, and Drum Separator
  - Dust Collector for Separator, Storage Hopper, and Blast Unit
  - Vacuum Recovery, Silo Tank, and Main Dust Collector
  - Dust Recovery, Supply Fan, Control Panel, and Dehumidifier
  - Gas Heater
- Painting Workshop Facilities:
  - Ventilation Fan, Supply Blower
  - Vacuum Recovery System
  - Dehumidifier: Independently operated for each painting bay
  - Heater: Independently operated for each painting bay

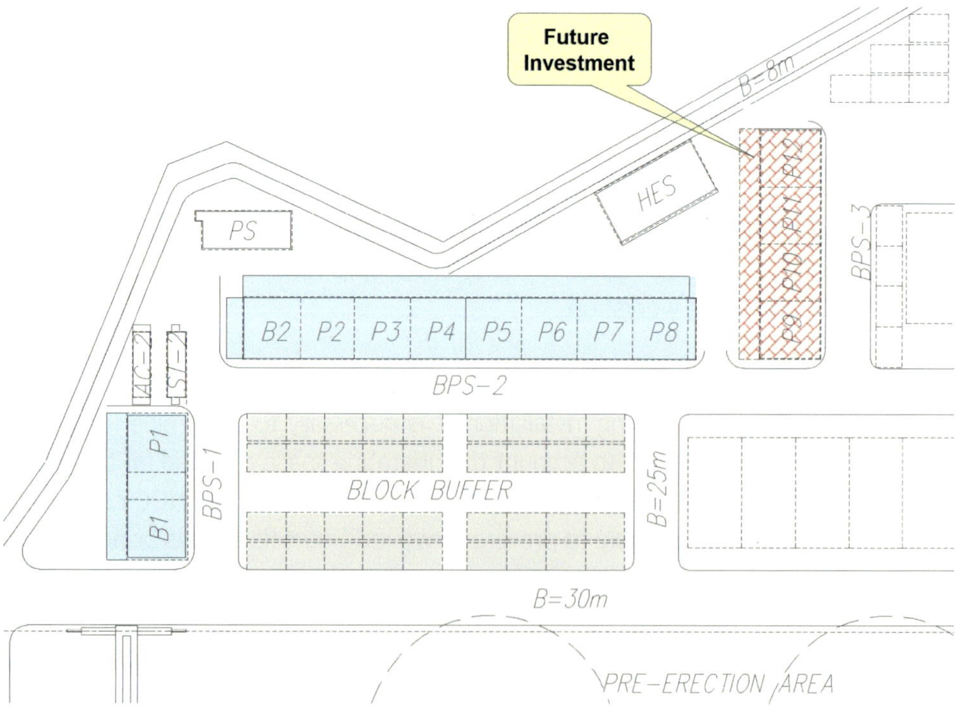

Figure 22 Typical Layout of Blasting and Painting Workshop (including Block Buffer)

3.13 Pre-Erection Area

In order to increase the annual turnover ratio of the dry dock, it is necessary to maximize the pre-erection while staying within the safe working load limit of the block erection crane. Therefore, a sufficient area should be secured around the dry dock for the pre-erection process.

First of all, it is highly desirable for the width of the pre-erection area, located on the side of the dry dock, to be comparable to that of the widest vessel in the product mix. The length of the remaining pre-erection area towards the dock head is calculated based on the total required area. Based on the typical arrangement of dry docks and pre-erection areas, as shown in Figure 23, the pre-erection areas should have a minimum required area that is equal to or greater than the area occupied by the dry dock floor of the ships being launched during the batch period of the dry dock (highlighted in pink). However, considering the difference in productivity between local shipyards and Korean shipyards, it is desirable to also include the occupied area (indicated in gray area) of ships under tandem construction. By referring to these guidelines and considering productivity variations caused by climatic conditions, it is possible to calculate the required pre-erection area, as indicated in Table 17.

Categorization	Dry Dock Operation	Lead Time for PE	Turnover Ratio of PE Area	Developed Area Ratio of PE	Required Area for Pre-Erection
Dock (Batch) Period	45 days	15 days for a single PE Block	3 times (=45-day/ 15-day) *(b)*	≈2.5 times of the Vertical Projected Area *(a)* (= Bottom Block PE + Main Deck Block PE + Deckhouse PE)	≈1.5 times of the Vertical Projected Area of Launching Vessel [= *(a)* / *(b)* / Productivity factor (67%) / Weather Impact (85%)]
Erection Quantity	45 times				
Erection Pitch Time	1 day/time				

Table 17 Required PE Area Assumptions

Meanwhile, since most of the pre-erection work is done vertically, the entire weight of the pre-erection block is transmitted as a concentrated load to the ground through a keel block, trestle, or stool. Therefore, a minimum bearing force of 30 tons/m^2 is required for safe operation in this working area.

- Pre-Erection Area Facilities:
  - Gantry Crane for PE: Refer to the dry dock section.
  - Jib or Travelling Tower Crane (TTC) for service: Refer to the dry dock section.
  - Block Lifter for Transporter PE: 8 sets of 350-ton lifting capacity. While not an essential facility, it is a highly valuable asset for pre-erection by using the transporters of LNGC cargo hold. The block lifter helps distribute the workload of the gantry crane during pre-erection and erection processes.

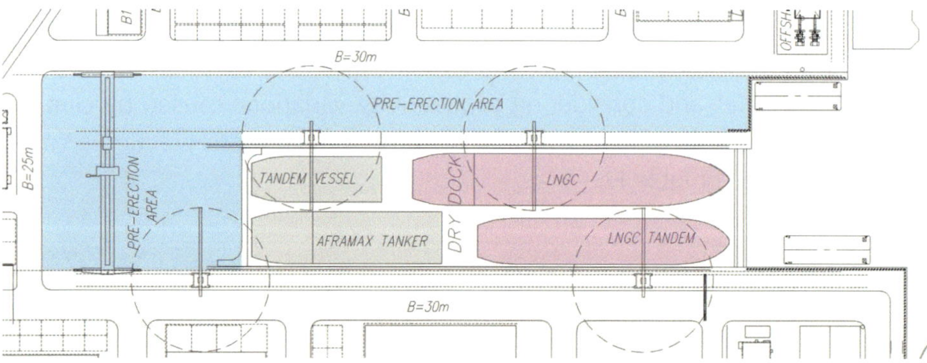

Figure 23 Typical Layout of Pre-Erection Area

3.14 Dry Dock

As for the type of shipbuilding berth that supports the keel of a ship, the traditional sliding berth is still commonly used in small and medium-sized shipyards today. However, for medium and large shipyards, three types of shipbuilding berths are typically used: dry docks, floating docks, and land-based construction areas. In this context, the following focuses on describing the most representative type of dry docks.

Dimension of Dry Dock

Based on the mid- and long-term production goals outlined in Section 2.1 of Part II, the requirements for tandem and parallel construction in dry docks should be reviewed. In the case of tandem construction, there is an advantage in maximizing the operational efficiency of the facility. This is because the workload of the large gantry crane for block erection can be maintained relatively evenly during the dock batch period. In the case of parallel construction, there is a disadvantage in that it increases the span of the large gantry crane for block erection. However, it allows for more effective adaptation to changes in the width of the ship being built, including general merchant ships and offshore products like super-large Floating Production Unit (FPU). Here, the contents describe the dimensions of the dry dock under conditions where tandem and parallel construction is possible.

Table 18 shows the standard dimensions for the vessel, as well as the minimum length requirements that will allow the half-ship to float safely without excessive trim for tandem constructions, without utilizing intermediate dock gates.

Type of Ship		Capacity	LOA	Width	Min. Tandem Length
Tanker	VLCC	320K DWT	333 m	60.0 m	177.0 m
	Suezmax COT	159K DWT	274 m	48.0 m	124.1 m
	Aframax COT	116K DWT	250 m	46.0 m	130.5 m
LNGC	Large Size	210K CBM	315 m	50.0 m	190.3 m
	Medium Size	173K CBM	290 m	45.8 m	157.5 m
	Conventional	151K CBM	291 m	42.4 m	162.5 m
Offshore	FPU (200K displacement)		341 m	63.8 m	187.5 m

Table 18 Referenced Dimensions by Vessel Type

Based on the dimensions mentioned above, it is necessary to determine the optimal dimensions by checking different combinations of the length and width of the dry dock. Tables 19 and 20 provide examples of the results.

For reference, in the case of a post-Panamax container ship (i.e., Length 366.5 m x Breadth 48.2 m) with a capacity of 13,000 TEU or more, where L/B is relatively large for a slender ship, a significant tandem length is required. Therefore, if an intermediate gate is not used, the tandem construction at the dry dock has some inappropriate aspects. Meanwhile, in Korean shipyards, there are many cases where these large container ships are newly built on floating docks.

Full-Ship	Tandem Ship		Space	Required Length
Suezmax COT L = 274 m	200K FPU	L=187.5 m	Total 12 m: - Hull to Dock Gate = min. 4 m - Hull to Hull = min. 4 m - Hull to Dock Head = min. 4 m	473.5 m
	210K LNGC	L=190.3 m		476.3 m
	173K LNGC	L=157.5 m		443.5 m
	320K VLCC	L=177.0 m		463.0 m
	159K COT	L=124.1 m		410.1 m
	116K COT	L=130.5 m		416.5 m
173K LNGC L = 290 m	200K FPU	L=187.5 m	Ditto	485.5 m
	210K LNGC	L=190.3 m		492.3 m
	173K LNGC	L=157.5 m		459.5 m
	320K VLCC	L=177.0 m		479.0 m
	159K COT	L=124.1 m		426.1 m
	116K COT	L=130.5 m		432.5 m
200K FPU L = 341 m	210K LNGC	L=190.3 m	Ditto	543.3 m
	173K LNGC	L=157.5 m		510.5 m
	320K VLCC	L=177.0 m		530.0 m
	159K COT	L=124.1 m		477.1 m
	116K COT	L=130.5 m		483.5 m

* *Note: The distance between the dock gate and the hull is based on a floating type of dock gate*

Table 19 Ship Length Combinations for Tandem Ship Construction Cases

Except for combinations of lengths that result in extreme outcomes and combinations that can lead to significant volatility in workload during the docking period, if the optimal value is selected (i.e., 483.5 m in this case), all types of ships can be built depending on the composition of the product mix during the relevant dock batch period.

On the other hand, when using an intermediate dock gate, additional space is required between the launch ship and the tandem ship. However, the tandem ship does not need to float, so the tandem length can be reduced. Therefore, the above value can be considered reasonable. Additionally, if the dock gate type is a hinge type rather than a floating type, the length of the dry dock must be increased by the depth of the dry dock.

A Line	B Line		Space	Required Width
Suezmax COT W = 48.0 m	200K FPU	W=63.8 m	Total Max. 12 m ~ Min. 4.5 m - Hull to Left Wall = 4.0~1.5 m - Hull to Hull = 4.0~1.5 m - Hull to Right Wall = 4.0~1.5 m	116.3 ~ 123.8 m
	210K LNGC	W=50.0 m		102.5 ~ 110.0 m
	173K LNGC	W=45.8 m		98.3 ~ 105.8 m
	320K VLCC	W=60.0 m		112.5 ~ ~~120.0~~ m
	159K COT	W=48.0 m		100.5 ~1 08.0 m
	116K COT	W=46.0 m		98.5 ~ 106.0 m
173K LNGC W = 45.8 m	200K FPU	W=63.8 m	Ditto	114.1 ~ ~~121.6~~ m
	210K LNGC	W=50.0 m		100.3 ~ 107.8 m
	173K LNGC	W=45.8 m		96.1 ~ 103.6 m
	320K VLCC	W=60.0 m		110.3 ~ 117.8 m
	116K COT	W=46.0 m		96.3 ~ 103.8 m
200K FPU W = 63.8 m	210K LNGC	W=50.0 m	Ditto	118.3 ~ 125.8 m
	320K VLCC	W=60.0 m		128.3 ~ 135.8 m
	116K COT	W=46.0 m		114.3 ~~~121.8~~ m

Table 20 Ship Width Combinations for Parallel Construction Cases

In terms of dry dock width, the following factors should be considered to determine the dry dock width: In parallel construction, where the widest vessels are placed in one row and the narrower vessels in the opposite row, a considerable number of combinations can be accommodated, except in a few extreme cases. In addition, the appropriate distance between the side wall of the dry dock and the side shell of the ship and the distance between ships must be considered in consideration of automatic welding of the block erection part, final painting work of the shell plate, and undocking of the launching ship. The recommended such working space is typically around 4 meters, but it is still possible to operate with as little as 1.5 meters of clearance.

Figure 24 shows examples of dry dock arrangements for tandem and parallel construction, based on the reviewed length and width of the dry dock.

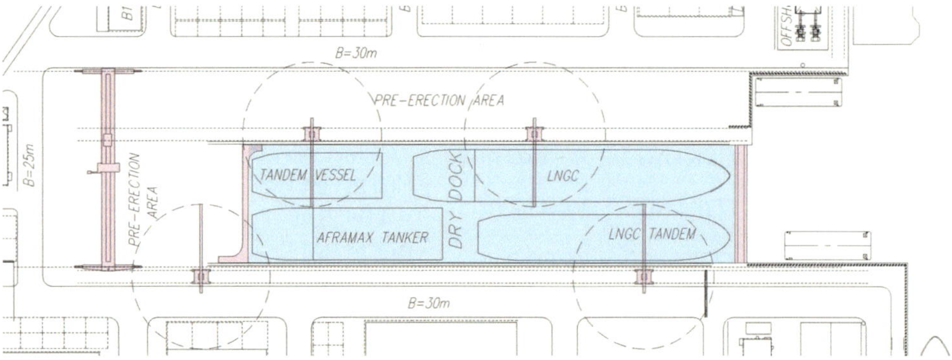

Figure 24 Typical Layout of Dry Dock

The depth of the dry dock can be determined by referring to Figure 25, which illustrates the sea level standard, and Table 21, which provides information on the draft of the launching ship and the tandem construction ship.

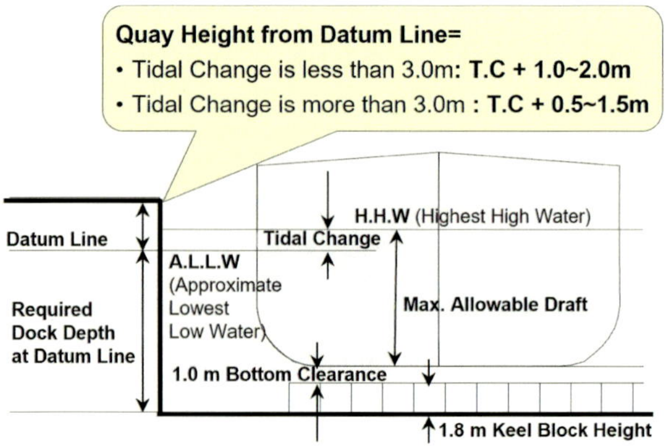

Figure 25 Required Depth of Dry Dock

Although the shipyard is located in an area protected by a breakwater or naturally sheltered from the open sea, if the magnitude of the tidal change at the proposed shipyard location is too high, it may not be suitable for shipyard construction or operation. Therefore, sea level tide information is an important factor to consider when selecting a shipyard site, and determining the depth of a dry dock starts from the Approximate Lowest Low Water (ALLW), which serves as the datum line for sea level. Although regulations regarding quay height may differ from country to country, if the tidal change is less than 3 meters, the quay height is determined by adding approximately 2 meters to account for sea level waves at the Highest High Water (HHW) level. The height of the keel block is set at 1.8 meters, which is the standard height for completing finishing work before launching, such as replacing timber on the keel block and painting. Bottom clearance refers to approximately 1 meter of additional space to accommodate appendages on the bottom of the launching ship, unexpected heeling when floating, or hull movement caused by waves.

Type of Ship		Capacity	Partial Floating Draft	Full-Ship Draft
Tanker	VLCC	320K DWT	6.68 m	Aft: 4.30m, Fwd: 3.77m
	Suezmax COT	159K DWT	6.35 m	Aft: 3.50m, Fwd: 3.03m
	Aframax COT	116K DWT	5.59 m	Aft: 3.50m, Fwd: 2.92m
LNGC	Large Size	210K CBM	6.0 m	Aft: 4.99m, Fwd: 4.71m
	Medium Size	173K CBM	5.74 m	Aft: 4.53m, Fwd: 4.48m
	Conventional	151K CBM	5.82 m	Aft: 4.58m, Fwd: 4.26m
Offshore	FPU (200K displacement) -			7.5 m

Table 21 Referenced Launching Draft by Type of Ship

Bearing Force on the Dry Dock Floor

The keel block arrangement on the floor of the dry dock is prepared by referring to the ship's lightship weight distribution curve, as shown in Figure 26. The weight of the ship is transferred to the bottom of the dry dock through the keel block, which is typically composed of concrete and wooden blocks. In particular, the arrangement of keel blocks after the engine room area is limited to the centerline of the keel due to the shape of the hullform. As a result, the concentrated load is inevitably applied to the longitudinal centerline of the ship. Therefore, the dry dock floor must have sufficient bearing capacity by dividing it into heavy and light zones.

Based on the keel block used in Korean shipyards, the required bearing capacity can be estimated as follows:

- Permissible load of the wooden keel block: 105 tons
- Area of the concrete keel block: 0.5 x 1.66 m = 0.83 m^2
- Uniformly distributed load: 105 tons / 0.83 m^2 = 127 tons/m^2

Another reason is that during the initial floatation process, the stern trim may cause relatively high concentrated loads. Therefore, it is necessary to establish a heavy zone with a high load capacity.

- Light Zone: 127 tons/m^2
- Heavy Zone: 152 tons/m^2 (= 127 x 1.2)

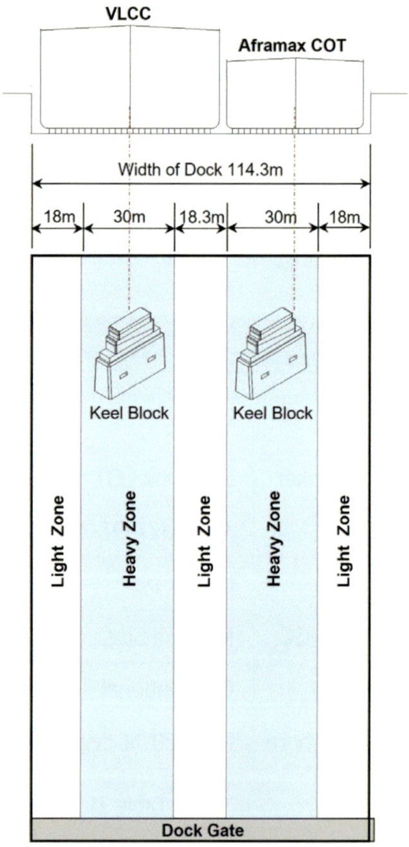

Figure 26 Heavy Zone on Dry Dock Floor

Cranes for Erection and Service

The large gantry crane, which is used to position blocks in the dry dock and pre-erection area, is commonly referred to as a Goliath crane. It is one of the key facilities that demonstrates the shipyard's production capa-city, and currently, there is a growing preference for cranes with

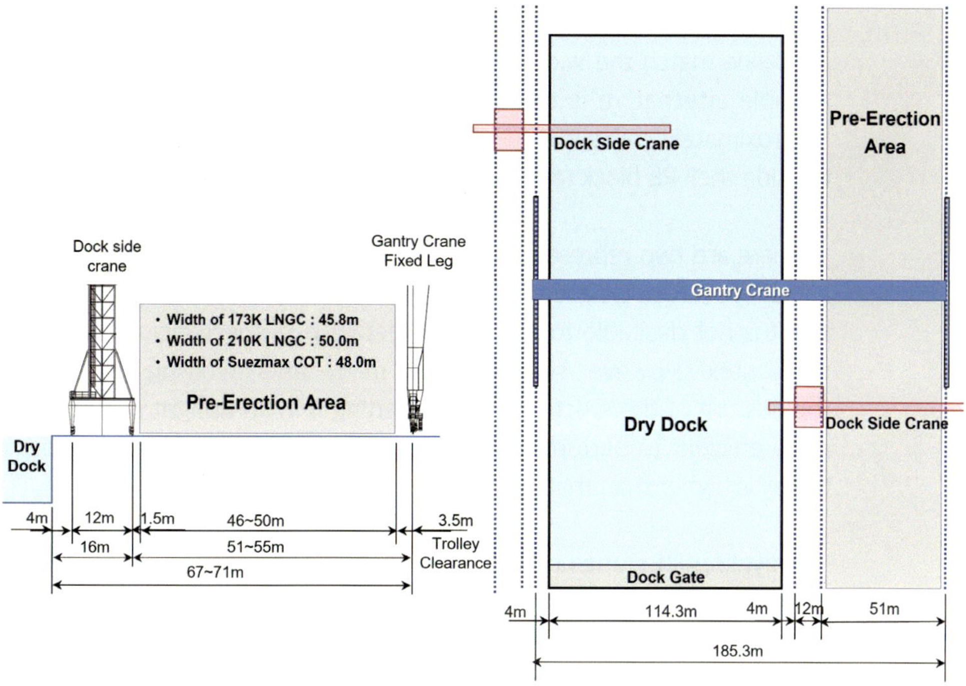

Figure 27 Typical Configuration of Gantry and Service Crane at Dock Side

high capacity. However, it is even more crucial to maintain a balance in the lead time between consecutive processes during the dock batch period, as well as the size of blocks produced in the assembly workshops.

Korean shipyards are focused on improving the operating ratio of this erection crane. With the addition of night shift work, the crane is able to perform 18 to 20 operations per day. It is also a good solution to install two cranes with the same capacity to increase the turnover ratio of the dry dock.

For large shipyards, the lifting capacity of erection cranes typically ranges from 900 tons to 1,500 tons, and above all, careful consideration is needed when determining the crane span. Depending on the span, it can have a significant impact on the weight and initial acquisition cost of the crane. It can also greatly affect the utilization of the dockside pre-erection area. Therefore, by referring to the contents and explanations depicted in Figure 27, the basic specifications of the erection crane can be determined.

It is ideal to have the effective width of the pre-erection area on the dock side match the width of the major ship type to be built. However, a viable alternative is to have the width of the main deck PE block (approximately ship's width - 2 x wing structure width) or the height of the side shell PE block (approximately Ship's depth) or more.

There are two representative types of gantry cranes: twin girder type and single girder type. Each type has its own operational characteristics, but it is not desirable to discuss the details of a specific crane model in this context. However, since there are many tasks involving turning over the posture of the erecting block or setting it in an upright position, it is more efficient to perform these tasks if the lifting capacity of the lower trolley is higher than that of the upper trolley.

Traveling jib cranes or traveling tower cranes are installed on both sides of the dry dock to serve as cranes used for installing equipment or various fittings on ships under construction in the dry dock and pre-erection area. The lifting capacity of these cranes naturally varies according to changes in their outreach, but the nominal capacities range between 60 tons and 200 tons.

- Dry Dock Facilities:
    - Gantry Crane for erection: 1 or 2 sets
    - Jib or Travelling Tower Crane (TTC) for service: 1 x 2 (Both sides of dry dock) set / (200 to 250 m)
    - Anti-Collision System for safe operation of cranes
    - Block Lifter for bottom block erection: 8 sets of 350-ton lifting capacity (This equipment is highly beneficial for improving the capacity utilization of the Goliath crane.)
    - Pump System for dry dock: 4 De-watering Pumps, 2 Stripping Pumps, 2 Drainage Pumps, 2 Ballast Pumps, 1 Auxiliary Ballast Pump, 2 Fire Pumps, 2 Fire Jockey Pumps
    - Dry Dock Accessories: Capstan, Bollard, Bitt, Fairleader, and more
    - Dock Gate: Floating type gate equipped with 2 sets of De-ballast Pumps
      (* *Note: In areas where seawater freezes during the winter season, it is necessary to have a de-icing system.*)

3.15 Land-Based Construction Area and Floating Dock

In the past, land-based construction was primarily used for small vessels like naval ships and special ships. However, the utilization of these production facilities has been further enhanced in the development of various offshore projects that require longer production lead times. Furthermore, even large merchant ships are being built recently at these land-based construction areas as an alternative when there is an excessive imbalance in the workload of offshore projects.

In the case of offshore projects, such as fixed platforms and topside modules, highly concentrated loads occur on the ground in the working area due to the unique shape of the structure, but it is very difficult to specify and limit load points due to the varying structure shapes depending on the project. Normally, the bearing force required for this working area is higher than the value of the pre-erection area. The required level of bearing force for each major working area in the shipyard will be addressed later.

For land-based products, the conventional method of load-out primarily involves the use of the grease skidding method with the gripper jack system. However, these days various new methods, such as the strand jack system and air pad sliding system, are being used. These methods offer advantages in terms of load-out weight and speed. On the other hand, when a ship is newly constructed at a land-based construction area, the position of the load point on the ground can be determined, allowing for faster load-out work using either a self-driving bogie system or a gripper jack with a roller skidding system.

As depicted in Figure 28, most offshore projects are transported using heavy cargo transportation vessels. However, when it comes to new shipbuilding, the launching process involves transferring the ship to a semi-submersible launching barge. This barge has significantly greater longitudinal strength compared to typical floating docks. Of course, in the case of small ships like naval vessels, there are instances where a synchronous lifting system is used instead of a semi-submersible launching barge for launching.

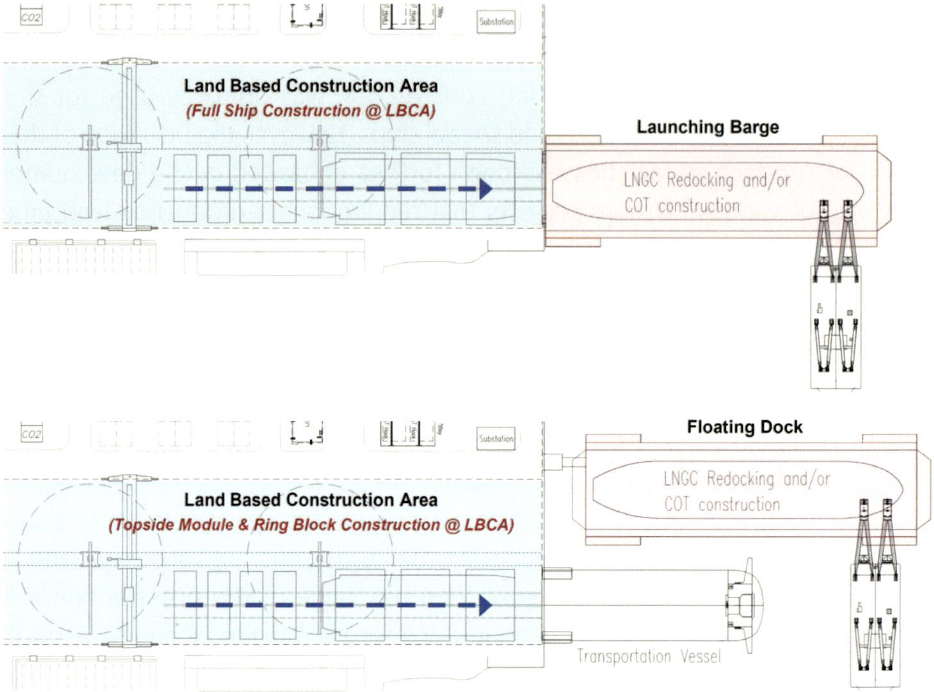

Figure 28 Typical Configurations of LBCA Construction

Meanwhile, certain Korean shipyards have adopted a shipbuilding method that involves pre-erecting sizable ring blocks that match the full width of the ship within a land-based construction area. Afterwards, the entire shipbuilding process can be completed by installing only about 10 of these large ring blocks on a floating dock using a floating crane.

The bearing capacity of the land-based construction area should be 30 to 50 tons/m^2, which is higher than the capacity required for the pre-erection area of the dry dock. For the construction of general cargo ships, when installing a bogie rail, it should be able to withstand a bearing force of 75 tons per wheel. On the other hand, it is desirable to apply the same concept as the dry dock to gantry cranes and service cranes. However, if the production volume is limited, mobile cranes, such as crawler cranes, can still be utilized, despite their somewhat inefficient productivity.

3.16 Quay

The outfitting quay is a production facility in the final stage of the shipbuilding process. In fact, it is very difficult to expand shipyard facilities according to additional requirements once the overall construction plan is finalized. Therefore, during the shipyard design stage, it is important to establish a shipyard development plan based on an analysis of future production goals and the potential for expansion. Additionally, since the duration of the quay varies greatly depending on the type of ship, it is necessary to take a reasonable approach to the construction of the primary types of ships to be built and future business plans. Table 22 provides an exemplary estimate of the quay berth requirements based on the quay period of Korean shipyards and the mid- and long-term production program that was reviewed previously.

<table>
<tr><th colspan="2">Categorization</th><th>1st~3rd Y</th><th>4th Y</th><th>5th Y</th><th>6th Y</th><th>7th Y</th><th>8th Y</th><th>9th Y</th><th>10th Y</th></tr>
<tr><td rowspan="3">Number of Launching Ship</td><td>COT</td><td>1</td><td>4</td><td>5</td><td>2</td><td>1</td><td>5</td><td>5</td><td>4</td></tr>
<tr><td>LNGC</td><td></td><td></td><td></td><td>3</td><td>3</td><td>4</td><td>4</td><td>4</td></tr>
<tr><td>FPU</td><td></td><td></td><td></td><td></td><td></td><td></td><td></td><td>1</td></tr>
<tr><td colspan="2">Quay Period in Korea</td><td colspan="8">• COT : 1.5~2.0 months, Container : 3.0± months,
• LNGC : 9± months,
• FPU : 26± months</td></tr>
<tr><td colspan="2">Assumptions at Local Site</td><td colspan="2">0.66 (≈1 ea) = [C5 x 2.0M x 1.59 / 24M]</td><td colspan="2">2.59 (≈3 ea) =[C7 x 2.0M x 1.39 / 24M] + [L3 x 9M x 1.59 / 24M]</td><td colspan="2">4.21 (≈5 ea) =[C6 x 2.0M x 1.13 / 24M] + [L7 x 9M x 1.39 / 24M]</td><td colspan="2">6.05 (≈6 ea) =[C9 x 2.0M x 1.05 / 24M] + [L8 x 9M x 1.18 / 24M] + [F1 x 26M x 1.59 / 24M]</td></tr>
<tr><td colspan="2">Productivity Factor based on Learning Curve</td><td>20.5%</td><td>12.4%</td><td>9.0%</td><td>7.1%</td><td>5.0%</td><td>4.1%</td><td>3.9%</td><td>3.7%</td></tr>
<tr><td colspan="2">Required Berth, except Unloading Berth</td><td colspan="2">1 ea</td><td colspan="2">3 ea</td><td colspan="2">5 ea</td><td colspan="2">6 ea</td></tr>
<tr><td colspan="2">Required Quantity for Double Mooring</td><td colspan="2">-</td><td colspan="2">-</td><td colspan="2">1</td><td colspan="2">2</td></tr>
</table>

Table 22 Required Berth Quantity of Outfitting Quay

In the table above, the quay period from launching to delivery assumes that productivity improves annually based on the learning curve. It starts at twice the difference compared to Korean shipyards, depending on the type of ship. Ships that exceed the maximum allowable number of quay berths at the shipyard, except for unloading quays for material logistics, must be double moored. This is illustrated in the exemplary quay layout shown in Figure 29. On the other hand, depending on the coastline of the shipyard site and the layout of the existing shipyard, there are many cases in which the number of mooring berths can be increased by adding a finger jetty to accommodate additional quays in the future.

The water depth of the quay must be sufficient to accommodate the adjustment of the draft during the sea trial and the inclining experiment prior to departure for the sea trial. There are no significant issues when it comes to a water depth of approximately 11 meters for large ships, except for the need to allocate space for installing large equipment such as retractable thrusters below the ship's baseline and floating dock operating areas.

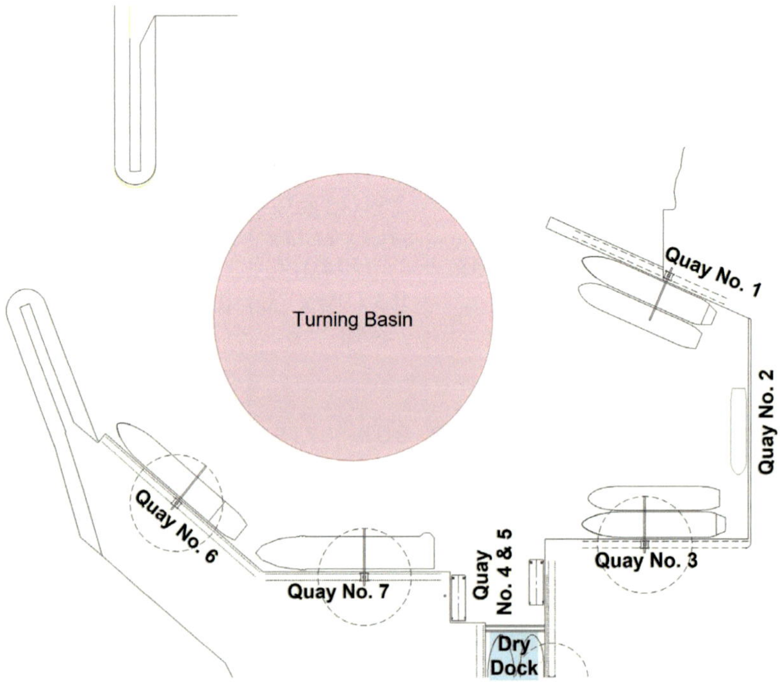

Figure 29 Typical Layout of Outfitting Quay

In addition, a turning basin is required for mooring a ship in a non-powered state at a quay or changing the direction of a ship when entering or departing from a sea trial. This turning basin must be located in a water area near the shipyard. On the premise that this operation can be safely performed with tugboats, the turning basin is designed to be twice the diameter of the vessel's Length Overall (LOA).

The quay service crane, which is a common production facility installed on the quay wall, needs to be a traveling crane capable of servicing the installation of fittings along the entire length of the moored vessel. Representative types of cranes include various kinds such as traveling jib cranes, Traveling Tower Crane (TTC), and Level Luffing Crane (LLC). When determining the capacity of this crane for a double mooring quay, it is important to carefully check and determine the lifting weight and outreach correlation based on the type of post-outfitting works. Figure 30 shows an example load diagram by crane outreach.

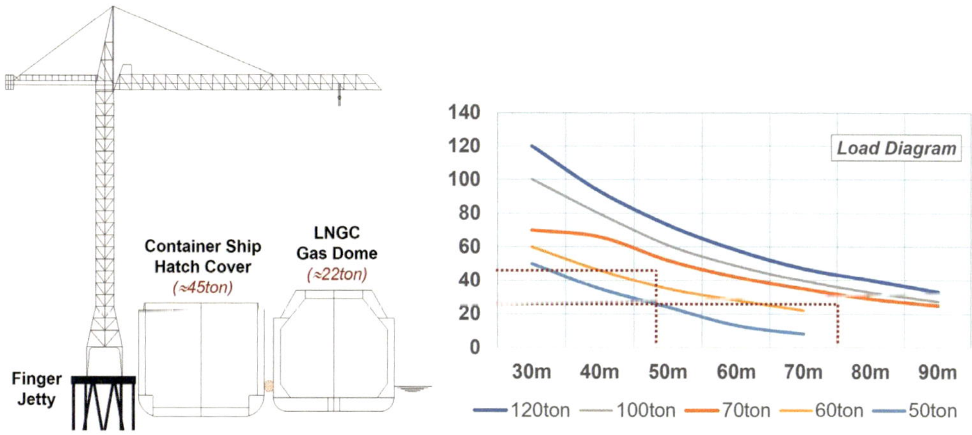

Figure 30 Lifting Capacity of Quay Service Crane

- Outfitting Quay Facilities:
  - Quay Service Crane: 1 set for a ship
  - Utilities (Electricity, Water, Compressed Air, Gases) feeding point: Every 25 meters distance

- Mooring Bitt: 50-ton Bitt (every 50 m distance), 200-ton Bitt prepared for tropical storm (6 ea in every mooring berth)
- De-icing System, if necessary

3.17 Machinery Workshop

Typical machined products in shipyards include propeller shafts, intermediate shafts, propellers, rudder stocks, and rudders. Generally, propellers and rudders are delivered in a finished state from specialized manufacturers. In the case of shafts and stocks, forged materials are primarily delivered and subsequently processed in the machinery workshop of the shipyard. Just before installation on the shipboard, manual surface machining is performed for the contact test between the propeller shaft and the propeller, as well as between the rudder and the rudder stock.

Table 23 categorizes the size and weight of products that require machining for large merchant cargo ships, and machinery workshop must be equipped with machining facilities that consider these product characteristics.

<table>
<tr><th>Machining Item</th><th>Dimension</th><th>Weight</th><th>Machining Process</th></tr>
<tr><td>Propeller Shaft</td><td>Diameter 580~670 mm
Length 7.9 ~ 13.3 m</td><td>25 ~ 36 tons</td><td rowspan="3">Centering → Roughing → UT → Finishing → MT→Boring → Honing 1/2 → Dimension Check → Propeller Contact (Lead time: 5 days) → Inspection</td></tr>
<tr><td>Intermediate Shaft</td><td>Diameter 490~575 mm
Length 1.2 ~ 12.4 m</td><td>4 ~ 26 tons</td></tr>
<tr><td>Propeller</td><td>Diameter 7.28 ~ 8.1 m</td><td>30 ~ 58 tons</td></tr>
<tr><td>Rudder</td><td>Width 6.5 ~ 8.2 m
Height 9.5 ~ 12.4 m</td><td>43 ~ 83 tons</td><td rowspan="2">Centering → Roughing → UT → Finishing → MT → Boring → Key & Deep Hole → Tiller Construction → Dimension Check → Fitting → 2nd Machining → Rudder Contact (Lead time: 5 days) → Inspection</td></tr>
<tr><td>Rudder Stock</td><td>Diameter 585~685 mm
Length 7.58 ~ 9.78 m</td><td>9 ~ 22 tons</td></tr>
</table>

∗ *Note: UT - Ultrasonic Test, MT - Magnetic Test*

Table 23 Physical Characteristics of Shaft, Propeller, and Rudder

In well-developed outsourcing environments for machining works, propeller shafts and rudder stocks may be supplied to shipyards in a state where the final machining is completed outside the shipyard. The only exception is the contact test, which is conducted using a large crane in the pre-erection area or dry dock.

- Machinery Workshop Facilities:
  - Lathe Machine
  - Vertical Turning Lathe Machine
  - Milling Machine
  - Plano Miller
  - Boring Machine
  - Grinding Machine
  - Hydraulic Cylinder Grinding Machine
  - Centering Machine
  - Drilling Machine
  - Radial Drilling Machine
  - Band Saw Machine
  - Slotting Machine
  - Overhead Crane
  - Equipment for the contacting process: Portable Boring Machine, Portable Taper Boring Machine, NC Taper Boring Machine, and Taper Boring Machine

3.18 Outfitting Manufacturing Workshops

In the past, shipyards used to manufacture various outfitting items in-house, including piping, ductwork, hull outfittings, and carpentry. Nowadays, shipyards tend to prioritize on-site work and outsource the manufacturing of outfittings that can be obtained externally. Nevertheless, the custom-made pipes, which are manufactured based on field measurements, must be manufactured in-house at the shipyard at the very least. These pipes are used for connecting sections of piping after the block has been pre-erected or erected. Additionally, it is desirable to manufacture pipes for pipe unitization work in order to facilitate efficient piping installation in narrow spaces.

Piping Manufacturing Workshop

As indicated in Table 24, there is significant variation in the quantity of pipe spools and the proportion of ferrous pipes to non-ferrous pipes based on the type of ship. In addition, the proportion of pipes made of special materials, which are relatively difficult to manufacture, is higher in offshore projects. Therefore, considering these characteristics and the shipyard's product mix strategy, it is necessary to define an appropriate production capacity for manufacturing piping spools in-house.

Categorization		Spool Quantity		Weight		In-House	Outsourcing
Ferrous Pipe	151K LNGC	12,919 ea	71%	529 tons	68%	6.3%	93.7%
	159K COT	6,215 ea	94%	460 tons	99.5%		
Non-Ferrous Pipe	151K LNGC	5,229 ea	29%	247 tons	29%	80%	20%
	159K COT	368 ea	6%	2.3 tons	0.5%		

Table 24 Ferrous and Non-Ferrous Pipes by Type of Ship

Based on the product mix strategy and the provided reference values for the number of pipe spools by ship type, the annual required number of pipe spools can be estimated. Furthermore, based on the shipyard's outsourcing policy, the shipyard itself can determine the monthly production quantity of pipe spools. The piping manufacturing workshop requires approximately 1.2 m^2 per pipe spool for the monthly production volume, and about 0.78 m^2 of outdoor buffer area for palletizing the manufactured pipe spool. Here, this index value includes space for pipe unitization work. Since this process is relatively simple, the difference in productivity can be overcome quickly, so it is not considered separately.

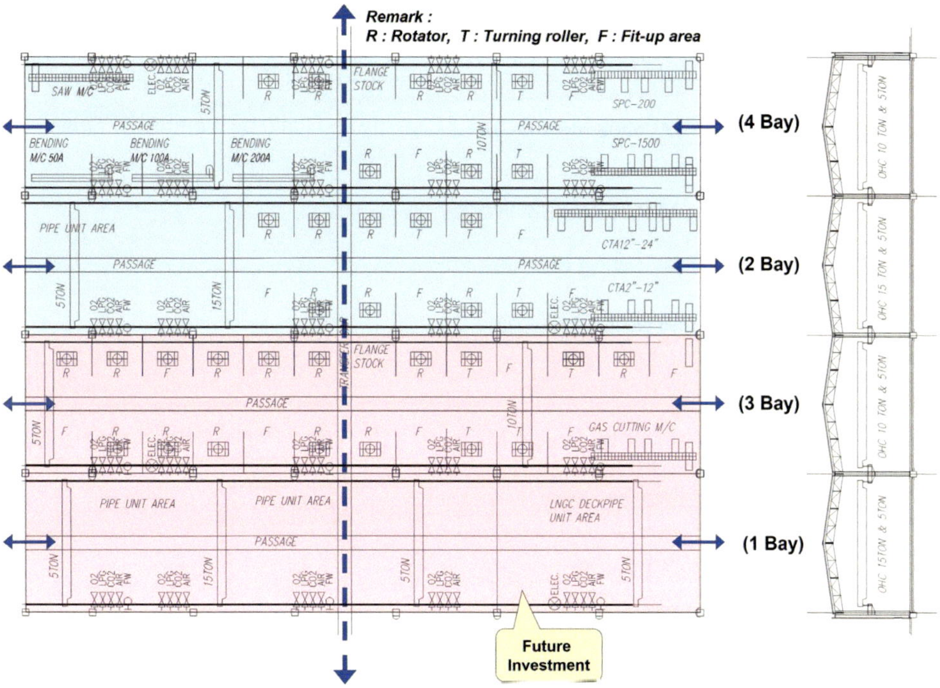

Figure 31 Typical Layout of a Piping Manufacturing Workshop

The layout of the piping manufacturing workshop depicted in Figure 31 is primarily divided into several areas. These include the pipe cutting and bending areas, the fit-up and tack welding workbenches for pipes and pipe fittings, the rotators and turning rollers for full welding, and the pipe unitization working area. It is necessary to separate the working area for steel pipes and non-ferrous pipes as much as possible in a pipe manufacturing workshop. The workbench configuration should be based on Table 25 below.

<table>
<tr><th>Categorization</th><th>Fit Up Workbench</th><th>Welding (at Rotator)</th><th>Welding (at Turning Roller)</th></tr>
<tr><td rowspan="2">Ratio</td><td>1</td><td colspan="2">2.5</td></tr>
<tr><td>-</td><td>11</td><td>4</td></tr>
<tr><td>Daily Capacity</td><td>-</td><td colspan="2">11.3 Spools</td></tr>
<tr><td rowspan="2">Workbench Quantity (ea)</td><td>-</td><td colspan="2">WB = Annual Sum of Spools / Working Day / 11.3</td></tr>
<tr><td>= WB / 2.5</td><td>= WB x 11 / (11 + 4)</td><td>= WB x 4 (11 + 4)</td></tr>
</table>

Table 25 Daily Piping Manufacturing Capacity and Ratio by Workbench Type

- Piping Manufacturing Workshop Facilities:
  - Gas Cutting Machine
  - Saw Cutting Machine
  - Beveling Machine
  - Horizontal Boring Machine
  - Lathe
  - Rotator
  - Turning Roller
  - TIG and CO_2 Welding Machines
  - Overhead Crane
  - Wall-Mounted Small Crane

Steel Outfitting Manufacturing Workshop

In the case of steel outfitting, it may not be necessary to have a production facility within the shipyard. However, if the outsourcing environment is not well-developed, it may be necessary to have a related facility within the shipyard or in a nearby location. Table 26 summarizes the quantity of steel outfitting by type, and the layout of the workshop is shown in Figure 32.

(Unit: Ton)	Hull Outfittings							Pipe Unitization	
	Mach. Seat	Hatch Cover	Traffic Fitting	Duct	Deckhouse Fitting	Anode Fitting	Electric Seat	Unit Fitting	Others
159K COT	44	30	213	46	10	0.2	33	739	184
	11.7%	7.9%	56.7%	12.2%	2.6%	0.1%	8.7%	-	-
173K LNGC	127	31	275	99	20	0.8	152	846	255
	18.0%	4.4%	39.0%	14.1%	2.8%	0.1%	21.5%	-	-

Table 26 Component Ratio of Steel Outfittings

- Steel Outfitting Manufacturing Workshop Facilities:
    - Cutting Machine
    - Angle Shearing Machine
    - Band Saw Machine
    - Knuckle Press
    - Hydraulic Press
    - Lathe
    - Drilling Machine
    - Hole Punch Machine
    - Auto Blasting Machine
    - Overhead Crane

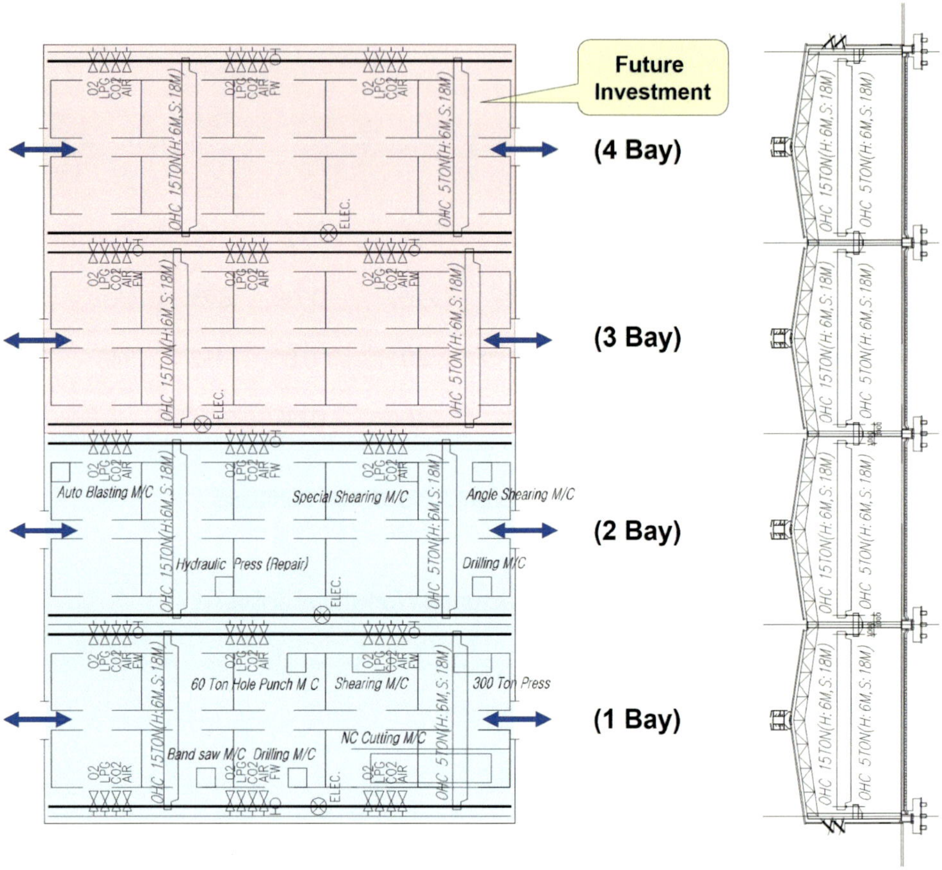

Figure 32 Typical Layout of Steel Outfitting Workshop

Galvanizing Workshop

Among piping materials and steel outfitting, the amount of plating, particularly galvanizing, required can reach a significant level, as demonstrated below. If the outsourcing environment is not well-developed near the shipyard, it may be necessary to have a galvanizing workshop on site or in a nearby location.

- 159K COT: Approximately 71 tons
- 173K LNGC: Approximately 155 tons

The galvanizing process generally involves several steps, including ① Heat Treatment for Degreasing, ② Pickling, ③ Rinsing and Water Cleaning, ④ Fluxing, ⑤ Drying, ⑥ Hot Dip Galvanizing, ⑦ Water Cooling and Chromate Treatment, ⑧ Inspection.
Shipyards should refer to these steps and organize related workshop accordingly.

In addition, stainless steel pipes often require a passivation process, so it is necessary to have facilities related to passivation located near the galvanizing workshop. It is also advisable to locate the galvanizing workshop near the piping manufacturing workshop, considering logistics. Figure 33 provides an example of a galvanizing workshop layout.

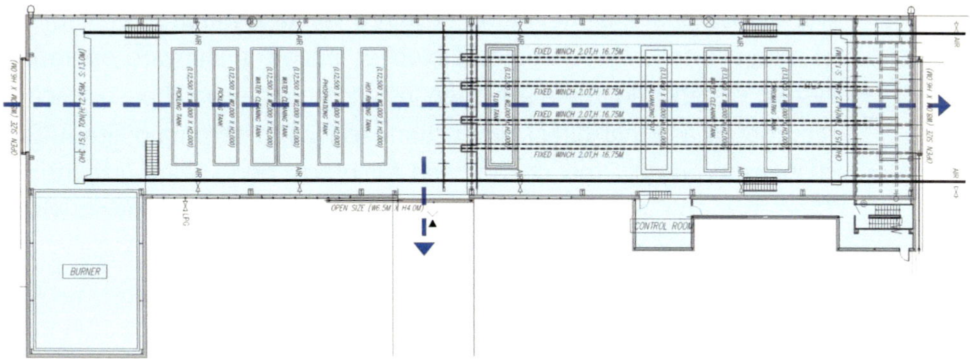

Figure 33 Typical Layout of Galvanizing Workshop

- Galvanizing Workshop Facilities:
    - Burner
    - Pickling Tank
    - Water Cleaning Tank
    - Phosphating Tank
    - Hot Rinsing Tank

- Flux Tank
- Galvanizing Tank
- Water Cooling Tank
- Chromating Tank
- Overhead Crane
- Hoist Crane

3.19 Warehouse and Materials Stock Area

In general, during the initial stage of shipyard operations, when the outsourcing environment is not well-developed, a relatively large warehouse space is inevitably required for the stable management of material delivery schedules. However, once the shipyard operating system is stabilized, there are often instances where warehouse space is repurposed as other direct production facilities to accommodate future changes in product mix or the need for production volume expansion. Therefore, this aspect should also be taken into consideration when making future expansion plans.

There are five types of warehouse spaces in shipyards.

- General Warehouse: For the storage of general goods
- Humidity and Temperature-Controlled Warehouse: For the storage of equipment sensitive to ambient conditions
- Flammable Liquid Storage: An independent warehouse for storing flammable substances such as paint and thinner
- Bonded Area and Warehouse: A bonded area for imported materials prior to customs clearance
- Outdoor Material Stock Area: An outdoor storage designated area for general stock items, including piping materials, steel fittings, and more.

In terms of utilizing the shipyard site, it is desirable to minimize the amount of warehouse space and allocate it to production facilities as much as possible. Therefore, it is crucial to minimize the waiting time from material warehousing to production input by effectively establishing Supply Chain Management (SCM) and Supplier Relationship Management (SRM) systems.

Meanwhile, fully automating warehouse space is not easy due to the diverse range of materials needed for shipbuilding. However, with the innovative development of information and communication technology, warehouses are now recognized as a field where automation technology can be applied, depending on the specific type of warehouse.

- Warehouse Facilities:
  - Overhead Crane
  - Forklift

3.20 Other Shipyard Production Facilities

There are several dedicated production facilities for LNGC shipbuilding, but they are not discussed in-depth here due to variations in production facilities based on the type of cargo containment system. These facilities include:

- Insulation Box and Panel Manufacturing Workshop
- Primary and Secondary Barrier Manufacturing Workshop
- Pump Mast Workshop
- Liquid Dome Box Workshop
- Pump Mast Pre-Outfitting Workshop
- LNGC Bottom Pre-Erection Area and System Scaffolding Assembly Area
- LNGC Dedicated Warehouse

<table>
<tr><th colspan="2">Facilities</th><th>Reference Data</th></tr>
<tr><td rowspan="2">Workshop</td><td>Cable Cutting</td><td>Cutting Machine Capacity:
1.5 Mil. Meters / machine,
(COT: 120,000 m, LNGC: 280,000 m)</td></tr>
<tr><td>Carpentry</td><td>Mostly Outsourcing</td></tr>
<tr><td rowspan="2">Maintenance</td><td>Production Facilities</td><td>For Spare Parts Storage and Maintenance</td></tr>
<tr><td>Heavy Vehicles</td><td>For Transporter and Forklift Maintenance</td></tr>
<tr><td rowspan="4">Building</td><td>Main Office</td><td rowspan="3">Min. Area:
6.5 (Canada) ~ 12 (UK) m^2/person</td></tr>
<tr><td>Site Office</td></tr>
<tr><td>Movable Office</td></tr>
<tr><td>Restaurant</td><td>Turnover Ratio: 2 ~ 2.6</td></tr>
<tr><td rowspan="6">Others</td><td>Fire Station</td><td rowspan="6">Depend on Local Regulations</td></tr>
<tr><td>Substation</td></tr>
<tr><td>Incinerator</td></tr>
<tr><td>Wastewater Treatment Plant</td></tr>
<tr><td>Parking Lot</td></tr>
<tr><td>Playground</td></tr>
</table>

Table 27 Other Facilities in the Shipyard

Other production facilities and indirect facilities are listed in Table 27. When designing a shipyard, it is recommended to refer to the country's design standards or practices.

3.21 Major Heavy Equipment in the Shipyard

Table 28 provides reference information for the requirements of various heavy equipment used in shipyards. These requirements are important factors to consider when estimating the initial Capital Expenditures (CAPEX).

<table>
<tr><th colspan="2">Shipyard Equipment</th><th>Reference Data</th></tr>
<tr><td>Floating Crane</td><td>3,000 ~ 8,000 tons Capacity of Floating Crane with Anchor Handling Boat</td><td>Use:
• New shipbuilding at a floating dock
• Installing topside module for offshore product
• Erection of LNGC bottom with system scaffolding in the dry dock</td></tr>
<tr><td>Tugboat</td><td>Bollard Pull Power:
≈ 56 tons</td><td>At least 3 tugboats are required for safely handling a large-sized vessel.</td></tr>
<tr><td rowspan="2">Transporter</td><td>Shipyard Transporter:
100, 200, 300, 350, 500, 600, 1,000 tons</td><td>For transporting a single block or a pre-erection block (Transporter coupling work for pre-erection blocks is equivalent to approximately 1.3% to 1.4% of the annual hull assembly weight.)</td></tr>
<tr><td>Self-Propelled Modular Transporter</td><td>For transporting topside modules, (Generally, rental equipment is used, and the required quantity is determined based on the shape and weight of the topside modules.)</td></tr>
<tr><td rowspan="2">Movable Crane</td><td>Crawler Crane</td><td>Temporary handling of heavy weight in places without permanent lifting facilities (Normally, rental equipment is used.)</td></tr>
<tr><td>Tire Crane:
30 ~ 300 tons</td><td>For loading and unloading equipment at a warehouse or material stock area</td></tr>
<tr><td>Fixed Crane (Movable)</td><td>Tower Crane:
Lifting 16 tons,
Outreach 50 meters</td><td>For pre-outfitting, tool and jig handling, and pallet handling in outdoor working areas (The tower crane's efficiency is better compared to using a mobile crane.)
• 1 set every 30 meters at the outdoor working area</td></tr>
</table>

Shipyard Equipment		Reference Data
For Worker Riding Vehicle	Cherry Picker: 15 ~ 47 meters	For working at a high position • 4 sets x annual number of delivery vessels
	Table Lift	For working at a high position • 1.3 set x annual number of delivery vessels
For Specific Tasks	Uni-Loader	For keel block handling at dry dock and floating dock
	Floor Sweeper	For collecting shot blasting balls
	Excavator equipped with magnetic device	For sorting cut pieces at the outdoor buffer area of the cutting workshop
For Logistics Work	Forklift Ratio = Engine Driven (4) : Battery Driven (1)	Engine Driven: 3 ~ 37 tons for outdoor Battery Driven: 1 ~ 5 tons for indoor • 5 sets x annual number of delivery vessels
	Tractor and Trailer	Tractor : Trailer = 1 : 3 ~ 4

Table 28 Reference Data for Shipyard Heavy Equipment

3.22 Required Capacity of Utilities

Electricity

Table 29 provides statistical reference data on power consumption by workplace in the shipyard. It is based on the assumption that the facility capacity of the steel stock yard is set to 1.0.

Shipyard Facilities	Machine Capacity (kw)	Load Factor (%)	Load Capacity (kw)
Steel Stock Yard	1.0	30%	0.3
Cutting Workshop	1.0	50%	0.5
Steel Treatment Workshop	1.3	56%	0.7
Panel Block Assembly Workshop	4.7	31%	1.4
Curved Block Assembly Workshop	1.3	31%	0.4
Blasting & Painting Workshop	9.3	31%	2.9
PE & Dry Dock Incl. Outdoor Working Area	13.3	31%	4.1
Goliath Crane	1.7	10%	0.2
Floating Dock	2.7	31%	0.8
Quay	3.3	31%	1.0
Air Compressor	18.7	31%	5.8
Office Building	1.7	50%	0.8
Others	6.7	25%	1.7
Total	67	-	21
Load Capacity / Total Machine Capacity	31%		
Incoming Electricity / Total Machine Capacity	45%		
Incoming Electricity	= 45% / 31% x Load Capacity		
Substation Distribution	Less than 300 meters distance between the substation and electric power panel		

Table 29 Electricity Consumption in the Shipyard

Gas, Compressed Air, and Water

Table 30 provides statistical reference data on the amount of compressed air, gases, and water required for various tasks in shipyards to produce 1.0 ton of assembly weight per year for different types of substances.

Substance	Unit	Reference Data
Carbon Dioxide	Ton	= 0.01743 x Annual weight of hull assembly
Oxygen	Ton	= 0.01369 x Annual weight of hull assembly
Ethylene	Ton	= 0.00094 x Annual weight of hull assembly
LPG	Ton	= 0.00387 x Annual weight of hull assembly
Argon	Ton	= 0.00123 x Annual weight of hull assembly
Nitrogen	Ton	= 0.00057 x Annual weight of hull assembly
Compressed Air	m^3	= 1,256 x Annual weight of hull assembly Consumption factor at mid-latitude site: Summer 100%, spring and autumn 80%, winter 60% * *Note: Excluding air consumption for de-icing system at sea*
Water	Ton	= 2.4 x Annual weight of hull assembly * *Note: Graywater is to be used for ballasting water on a tandem construction vessel.*

Table 30 Annual Consumption by Type of Substance *(Gases, Compressed Air, and Water)*

4 Shipyard Layout

4.1 Layout of Shipyard

Prior to designing the shipyard layout, it can be beneficial to start by identifying the characteristics of each parent shipyard. It can be a good reference to look at the layouts of representative shipyards in Korea, China, and Japan, which are solidifying their global position as leaders in the shipbuilding industry today. By carefully examining the aerial photos of representative shipyards in the three countries mentioned above, as shown in Figures 34 to 42 of 'Google Earth', it is possible to establish design guidelines that are suitable for the new shipyard site. Depending on the location and shape of the shipyard site, it can be roughly divided into two categories. First, this is a case where shipyard facilities are effectively arranged by utilizing the natural topography. In the second case, when a rectangular shipyard site is newly developed, the production facilities are then placed on this site. All Chinese shipyards presented here as examples are located downstream of rivers and share this latter characteristic. Of course, there is no absolute standard or correct answer for the specific layout of a shipyard. However, it is more desirable to derive an optimal layout plan through a trade-off study of various alternatives after examining the cases of different shipyards.

Key essential points to be reviewed in this process include the following:

Type, Location, and Quantity of Shipbuilding Docks

Based on the topography and geological conditions of the shipyard development site, it is important to determine the optimal location for the dry dock first. Additionally, it should be situated in a location where soil and sand do not accumulate around it as a result of ocean currents, tides, river flows, and other factors.

The size and quantity of shipbuilding dry docks should be determined by considering the advantages and disadvantages of each type of dock, and several types can be combined as needed.

- Single dry dock to build only one vessel
- Tandem construction dry dock that builds a full-ship and a half-ship

- Tandem and parallel construction dry dock that builds two full-ships and two half-ships
- LBCA (Land-Based Construction Area) that builds full-ships and/or offshore platforms without requiring a dry dock
- Floating dock to build a full-ship in association with a floating crane.

On the other hand, the combination of different types of shipbuilding docks should be determined based on the production goals set for the mid- and long-term. In particular, when it comes to land-based construction, there is the advantage of flexibility in production. Ships can be built in parallel based on the workload, alongside the production of offshore fixed platforms and topside modules. In the case of a floating dock, although the Build Method can be a bit challenging, the initial investment is relatively small compared to other types of shipbuilding docks. Additionally, it offers the flexibility to sell the fixed asset at any time.

Optimization of Logistics Flow

Relevant workshops and buffer areas must be carefully arranged to prevent reverse logistics from occurring in the flow of main production activities, starting from material unloading to block erection.

Of course, it may be desirable to implement line production as much as possible in order to eliminate wasteful elements that occur in the logistics flow. However, the limited feasibility of the JIT (Just-In-Time) production method in the shipbuilding industry is due to the constraints of product standardization and the limited number of ships constructed with the same design. If the entire assembly process, from steel cutting to fabrication, minor assembly, sub-assembly, and grand assembly, is line-manufactured in a single assembly workshop, several factors such as product change, introduction of new production equipment, changes in Build Method, and productivity improvements can continuously emerge new bottleneck processes. This may result in restrictions on the production capacity of a single assembly workshop operating in a line production system. Therefore, careful attention must be taken to configure the entire assembly process in a line production system, unless it is a shipyard that exclusively constructs a single type of ship.

Additionally, densely arranging production workshops without an appropriate buffer area based on the characteristics of the working process can lead to chronic traffic congestion. This congestion can be caused by the transportation of large blocks or slow speeds of transporters. Therefore, considering the output generated by major production workshops, the effective width of the main production roads is set according to the following standards:

- The production road from the unloading quay to the steel fabricating workshops should have a width for cross traffic of vehicles transporting raw materials, such as 12+α meters.
- The production road from the assembly workshops to the dry dock should have a width greater than the maximum size of the unit block determined by the production capacity of the assembly workshop, such as 22+αmeters.

On the other hand, interference in the flow of logistics can be minimized by distributing production support facility spaces, such as scaffolding and jig stock areas, and various warehouse spaces, as widely as possible throughout the shipyard. Additionally, it is desirable to minimize interference with production roads by establishing a road network at the outermost part of the shipyard boundary as much as possible. This outer road will serve as production support roads for warehousing equipment transported to the shipyard and for the movement of personnel within the shipyard. The width of these production support roads should be allocated to a width greater than or equal to the standard for general roads.

Bottleneck Process and Buffer Area

As mentioned earlier, unless only one product is being produced, it is natural for a new bottleneck process to appear in the main production process over time due to various factors. This bottleneck process can be permanent, but it is also very common for it to appear as a temporary phenomenon due to product changes, seasonal effects, and other factors. Therefore, when considering shipyard layout design, it is essential to allocate a certain level of buffer area, including waiting areas, based on the workshop's work characteristics. This allocation aims to maximize the workshop's operational efficiency, as it is a permanent facility, and minimize the potential disruptions caused by unexpected bottleneck processes. In particular, buffer areas are allocated around the workshop for the completion of non-essential work that can be done outside of the workshop. The appropriate level of buffer area was previously mentioned in Part II, Chapter 3: Capacity of Each Production Process.

- Buffer area for identifying and palletizing after steel cutting and fabrication
- Waiting area before being handed over to the subsequent process after minor assembling and sub-assembling
- Waiting area before transferring to the subsequent process after T-Bar manufacturing
- Buffer area for finishing work of pre-outfitting and inspection after taking the block out from the grand assembly workshop
- Buffer area for pre-outfitting work outdoors around the pre-outfitting workshop
- Buffer area for palletizing after pipe manufacturing and galvanizing
- Waiting area for blasting and painting, as well as a buffer area for outdoor drying after painting

In the process of expanding the production capacity of a shipyard, bottlenecks often arise in the buffer area, so it is essential to incorporate area planning into the baseline planning process.

Furthermore, allocating detailed location information for all production workshops, as well as individual block buffer areas, will be able to facilitate the transportation of blocks by autonomous transporters in the near future.

Future Expandability of Production Capacity

During the initial planning stage of shipyard construction, it is reasonable to establish a phased development plan. This is because developing the entire shipyard site at once requires a substantial initial investment and also requires a significant amount of time to become a stable shipyard. Generally, when profitability improves through continuous productivity innovation in the shipyard operating process, most executives naturally seek to expand production capacity by increasing the number of ships being built.

After the overall layout of the shipyard is completed, it may be difficult to expand essential production facilities like the dry dock and outfitting quay in the future. Therefore, even if many modifications are needed in the future due to the uncertain market environment, it is important to consider the expandability of future production facilities in the early stages of shipyard design, taking into account the future business portfolio.

On the other hand, if shipyard modernization or facility expansion is required within the limited shipyard site boundaries, only essential work processes that must be performed exclusively at the shipyard should be retained. Non-core production facilities, such as steel stock yards, outfitting manufacturing workshops, galvanizing workshops, and welfare facilities, can be outsourced or relocated outside the shipyard as an alternative.

Integration with Shipyard Construction Plan

As shown below, it is essential to establish a comprehensive project master plan for the entire project, considering that it takes a considerable amount of construction timeline from shipyard design to the completion of major construction. In order to minimize the construction period, if construction priorities can be managed so that various steel structures such as workshop steel structures, dry dock gates, and Goliath cranes required for shipyard construction can be produced in-house, not only the construction period but also the construction budget can be very effectively reduced. To support this, it is effective to make a material unloading quay available in the early stages of construction, and also to make some bays of the curved block workshop available for temporary steel cutting, fabrication, and assembly work. Additionally, allowing the pre-erection area in front of the dry dock head to be constructed early allows the Goliath cranes to be ready for final assembly and erection. In addition, it is desirable to establish a construction plan that enables the erection of the dock gate and floating dock to be carried out as much as possible in parallel with the progress of the dry dock floor construction.

- Dry dock : 30 ± months
- Floating dock : 20 ± months
- Goliath crane : 20 ± months

A Place Where Ships are Built

Figure 34 Korean Shipyard - DSME (currently Hanwha Ocean) *(Google Earth Photo)*

Figure 35 Korean Shipyard - HHI *(Google Earth Photo)*

지도(제목 없음)
Google Earth

Figure 36 **Korean Shipyard - SHI** *(Google Earth Photo)*

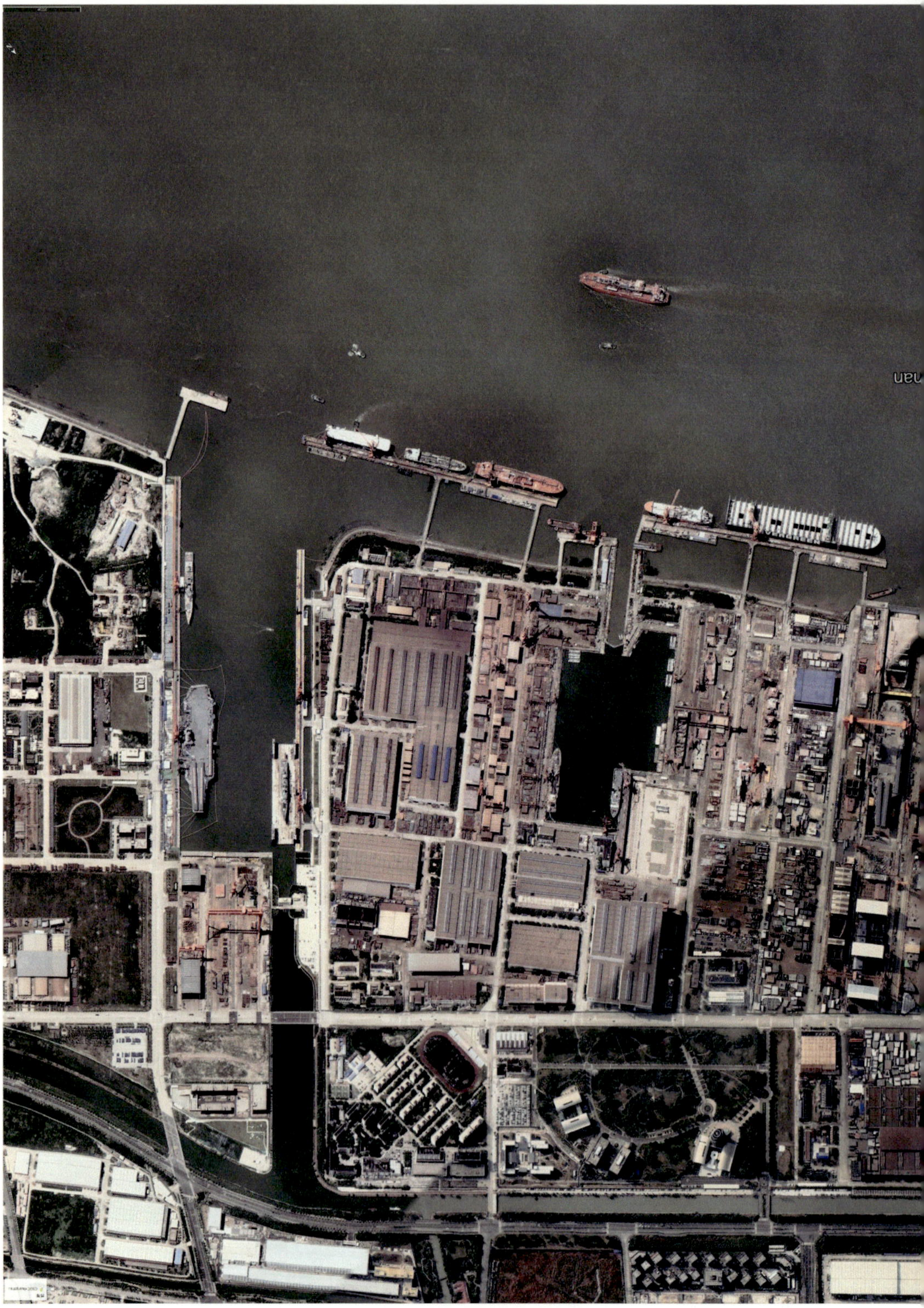

Figure 37 **Chinese Shipyard - CSSC Jiangnan** *(Google Earth Photo)*

Google Earth

Figure 38 Chinese Shipyard - CSSC SWS *(Google Earth Photo)*

Google Earth

Figure 39 Chinese Shipyard - CSSC Guangzhou *(Google Earth Photo)*

Imabari Shipb

Figure 40 Japanese Shipyard - Imabari Shipbuilding *(Google Earth Photo)*

JMU

Figure 41 Japanese Shipyard - JMU *(Google Earth Photo)*

Yokosuka Works of Sumitomo Heavy Industries, Ltd.
Sumitomo Heavy Industries, Yokosuka Works

Figure 42 Japanese Shipyard - Sumitomo Shipbuilding *(Google Earth Photo)*

4.2 Area Ratio

Before finalizing the shipyard layout, it can be meaningful to calculate the area ratio for each major section and compare it with the parent shipyard to ensure a balance in the layout design. It should be noted, for reference, that Table 31 indicates the area ratios for each major section of a shipyard in a mid-latitude region with favorable climatic conditions. However, it is important to understand that these ratios can vary significantly based on different climatic conditions.

Workshop and Area	Ratio (%)	Remarks
Steel Fabrication Workshop	4.1	Covered Workshop Area: 15.6% (Depending on the climate conditions, it can exhibit significant variations.)
Hull Assembly Workshop	6.0	
Pre-Outfitting Workshop	1.5	
Blasting & Painting Workshop	2.4	
Piping Manufacturing Workshop	0.4	
Deckhouse Assembly Workshop	0.2	
LNGC Dedicated Workshop	1.4	
Pre-Erection Area	10.7	
Dock Area	6.5	
Block Inspection Area	1.2	
Block Buffer Area	9.3	
Quay Area	3.8	
Warehouse	1.0	
Buildings	1.3	
Other Work Areas	2.0	
Production Support Area	7.8	
Road	16.0	
Other Small Area	24.4	
Total	100.0	

Table 31 Area Ratio of Shipyard Layout

4.3 Bearing Force at Each Workshop

As shown in Table 32 below, the required bearing capacity varies for each shipyard workplace, so the shipyard layout must be carefully considered to minimize the need for excessive ground reinforcement construction when designing the shipyard layout.

<table>
<tr><th colspan="2">Workshop Area</th><th>Bearing Force (t/m^2)</th><th>Remarks</th></tr>
<tr><td rowspan="2">Steel Stock Yard</td><td>Steel Stock Area</td><td>10</td><td>Gravelly Field</td></tr>
<tr><td>Pre-Assembling Plate Area</td><td>5</td><td>A part of the Steel Stock Yard</td></tr>
<tr><td colspan="2">Steel Treatment Workshop</td><td>5</td><td></td></tr>
<tr><td rowspan="2">Cutting Workshop</td><td>Transfer Bay</td><td>10</td><td rowspan="2"></td></tr>
<tr><td>Other Area</td><td>5</td></tr>
<tr><td colspan="2">T-Bar Workshop</td><td>5</td><td></td></tr>
<tr><td rowspan="2">Bending Workshop</td><td>Under Machine Area</td><td>Variable</td><td rowspan="2"></td></tr>
<tr><td>Other area</td><td>5</td></tr>
<tr><td rowspan="3">Panel Block Workshop</td><td>Bay 1</td><td>10</td><td rowspan="3">The bearing capacity under workshop columns is determined by the crane's lifting capacity.</td></tr>
<tr><td>Bay 2 ~ 4</td><td>5</td></tr>
<tr><td>Bay 5 ~ 7</td><td>10</td></tr>
<tr><td rowspan="2">Curved Block Workshop</td><td>Bay 1</td><td>5</td><td rowspan="2">Ditto</td></tr>
<tr><td>Bay 2 ~ 3</td><td>10</td></tr>
<tr><td colspan="2">Block Inspection Area</td><td>10</td><td></td></tr>
<tr><td colspan="2">Deckhouse Assembly Workshop</td><td>10</td><td></td></tr>
<tr><td colspan="2">Pre-Outfitting Workshop</td><td>10</td><td></td></tr>
<tr><td colspan="2">Blasting and Painting Workshop</td><td>10</td><td></td></tr>
<tr><td rowspan="2">Machinery Workshop</td><td>Under Machine Area</td><td>Variable</td><td rowspan="2"></td></tr>
<tr><td>Other Area</td><td>10</td></tr>
</table>

Workshop Area		Bearing Force (t/m²)	Remarks
Other Workshop	Piping Manufacturing WS	5	WS: Workshop
	Steel Outfitting Workshop	5	
	Carpentry Workshop	5	
	Pump Tower PE Workshop	10	For LNGC
Outdoor Working Area	Pre-Erection Area	30	
	System Scaffolding Area	5	For LNGC
	Block Buffer	5	Partial differential settlement allowed.
Dry Dock	Heavy Zone	152	Under the hinged leg of the Goliath crane: 80 tons/m²
	Light Zone	127	
	Goliath Crane Rail	90	
	Service Crane Rail	35	
Land-Based Construction Area		30 ~ 50	For construction of the offshore product
Outfitting Quay	Mooring Quay	5	
	Skidding Wheel Load	75	A part of LBCA
Road	Primary Production Road	10	For block transportation
	Other Road	5	

Table 32 Bearing Force at Each Workshop

5 Shipyard Equipment and Facilities

5.1 Investment List by Shipyard Development Stage

In Chapter 3 of Part 2, the equipment required for each workshop for each individual production process is listed. In order to calculate CAPEX at each stage of shipyard development, it is necessary to categorize investment items by workspace according to the investment stage, as shown in Table 33 below. Furthermore, it is necessary to put together a brief outline of the specifications for the equipment and facilities to be included in this list.

Based on this list, a typical procurement process for shipyard equipment would involve initially going through a pre-qualification process with multiple shipyard equipment suppliers, evaluating their proposals for technical specifications and price, and ultimately confirming the final purchase order specifications.

- Steel Stock Yard
- Steel Treatment Workshop
- Cutting Workshop
- T-Bar Workshop
- Bending Workshop
- Panel Block Workshop
- Curved Block Workshop
- Deckhouse Assembly Workshop
- Pre-Outfitting Workshop
- Blasting and Painting Workshop
- Pre-Erection Area, Dry Dock, LBCA
- Launching Barge and/or Floating Dock
- Outfitting Quay
- Machinery Workshop
- Outfitting Manufacturing Workshops
- Warehouse
- Other Shipyard Facilities, including major heavy equipment

Meanwhile, it is challenging to generate easily referenced data for the initial investment cost calculation due to the various cost variables that need to be considered by country. However, by referencing Table 34, which indexes the proportion of each major cost account, the feasibility can be objectively evaluated by comparing the initial acquisition cost of a large shipyard and the acquisition cost of new assets installed during shipyard operation with the CAPEX of a new shipyard.

Workshop	Bay	Equipment and Facilities	Brief Specifications	Quantity		
				1st	2nd	Sum
Cutting Workshop	Bay 3	CNC Plasma Cutting Machine	Rail span: 11.5 m Two Plasma Torch Stations Effective Cutting Width: 10.5 m (workbench size) Effective Cutting Length: 81 m (workbench size) Allowable Cutting Thickness: - Plasma Torch: Ferrous 6~40 mm Cutting Speed: - Plasma Torch: 0 ~ 6,000 mm/min - Rapid Speed: Faster than 24 m/min Marking Speed: Faster than 24 m/min	2		2
		CNC Plasma Cutting Machine	One Plasma Torch Station		1	1
		25/5 tons Magnetic Overhead Crane	Rail Height: 9.1 m, Rail Span: 28.0 m Motor Speed: Main Hoist 10 m/min - Aux. Hoist 5 tons Monorail Hoist (under girder) - Traversing 30 m/min - Travelling 100 m/min - Beam Rotating 1 rpm Operation Method: Fixed Cabin (under trolley)	1		1
		3 tons Semi-Gantry Magnetic Crane	Rail Height: 4.5 m, Rail Span: 19.0 m	1		1

Table 33 Sample List of Shipyard Equipment and Facilities by Development Stage

Categorization		Ratio	Remarks
Land	Shipyard Area	71.7%	
	Residential Area	14.2%	Outside the Shipyard
	Supplementary Facilities Area	3.4%	
	R&D and Training Institute	10.7%	Another Province
	Subtotal	33.9%	
Building	Production Facilities	50.3%	Incl. Workshop, Substation
	Office Building	22.9%	
	Welfare Facilities	6.5%	Incl. Cafeteria, Dormitory
	Supplementary Facilities	20.3%	Incl. Apartment
	Subtotal	17.9%	
Nonbuilding Structures	Dry Dock & Quay	78.9%	
	Pavement & Utilities	11.1%	
	Outdoor Working Area	6.4%	Incl. Temporary Shelter
	Auxiliaries	0.9%	Incl. Fence, Lighting Tower
	Supplement Facilities	0.9%	Incl. Drainage & Septic Tank
	Landscape Facilities	0.9%	
	Subtotal	19.6%	
Vehicle and Transport Equipment	Construction Equipment	13.5%	Incl. Mobile Crane, Forklift
	Special Transport Equipment	25.7%	Incl. Transporter, Cherry Picker
	Cargo Truck	2.0%	
	Car and Van	1.4%	
	Other Vehicles	0.9%	
	Vessel	52.6%	Incl. Tugboat, Floating Crane
	Aircraft	3.8%	Helicopter
	Subtotal	7.3%	
Mechanical Equipment	Computation Equipment	0.1%	POS System
	Experimental & Research EQPT	2.8%	EQPT: Equipment

Categorization		Ratio	Remarks
Mechanical Equipmen	Crane	43.4%	
	Welding Machine	9.0%	Incl. Gouging Machine
	Environmental Facilities	0.5%	Incl. Dust Collector
	Surface Treatment Facilities	0.3%	
	Production Support Facilities	11.8%	Incl. Air Compressor, Cooler
	Steel Fabrication Equipment	2.8%	Incl. Roll Press, Bend Machine
	Cutting Machine	4.1%	
	Machinery and Equipment	1.9%	Incl. Lathe, Drilling Machine
	Heating and Cooling Equipment	0.3%	
	Blasting and Painting Equipment	8.8%	
	Transferring Equipment	6.6%	Incl. Conveyor, Block Lifter
	LNGC Dedicated Equipment	1.0%	Incl. Invar Fabrication EQPT
	Carpentry Equipment	4.6%	
	Other Equipment	2.0%	Incl. Pipe Rotator, Fan, Pump
	Subtotal	14.5%	
Tool & Jig	Jig	85.0%	Incl. Trestle, Support Tower
	Measuring Tool	3.6%	
	Common Tool	11.4%	Incl. Chain Block, Level Gauge
	Subtotal	3.9%	
Fixtures	Office Equipment	12.9%	
	Computerized Equipment	53.6%	Incl. Laptop Computer
	Maintenance Equipment	4.6%	
	Communication and Optic EQPT	13.4%	
	Common Appliances	2.8%	
	Household Appliances	3.8%	
	Kitchen Appliance	9.0%	
	Subtotal	3.0%	
Total		**100.0%**	

Table 34 Proportion of CAPEX by Account Item

5.2 Manufacturer Pre-Qualification and Technical Specifications

Shipyard production facilities have to assemble large and heavy objects, which are divided into numerous blocks one by one, from super-large hull structures. Therefore, it is crucial to ensure that the performance, reliability, and durability of the product are sufficiently verified. Additionally, in order to promote continuous production activities and enhance productivity innovation, it is crucial for the production facility to have excellent maintainability. This will enable workers who are directly involved in production to easily carry out Total Productive Maintenance (TPM) activities. While there may be constraints on the initial acquisition budget, it is important to consider the operating costs incurred during the facility's life cycle.

Therefore, it is advisable to prepare a technical specification for the final purchase through comparative evaluation after receiving preliminary price quotations and technical specifications from several reputable international manufacturers. If a manufacturer that is not highly recognized is included, it is necessary to proceed with a pre-qualification process for that manufacturer. At this stage, it is helpful to pre-determine the weights of different evaluation factors, such as price, performance, durability, maintainability, and sales performance, to facilitate a rational equipment selection process.

Although there will be standardized templates for each company, the technical specification generally includes the contents expressed in italics below and requires careful attention from the facility introduction stage.

<u>Contents of Technical Specification</u>

I. General Provisions

1. *Scope of Specification*
2. *Conflicting Requirements*
3. *Technical Variations*
4. *Tender's Obligations for Coordination with the Employer*
5. *Tender's Responsibilities*

6. *Design and Manufacturing*
7. *Material and Workmanship*
8. *Facilities for the Representative*
9. *Subcontracting*
10. *Project Master Scheduler and Monthly Progress Report*
11. *Quality Assurance Requirements*
12. *Quality Control by the Tender*
13. *Inspection and Testing*
14. *Delivery Date and Delivery*
15. *Painting and Protective Coating*
16. *Preparation for Shipment and Preservation*
17. *Delivery Terms and Conditions*
18. *Warranty*
19. *Worldwide Service Network*

II. Scope of Work

1. *Overview*
2. *Scope of Supply*
3. *Documentation and Manuals*
4. *Engineering Documents*
5. *Spare Parts and Special Tools*
6. *Supervision and Training Program*

III. Technical Specification

1. *General*
2. *Design and Manufacturing Criteria*
3. *Design Requirements*
4. *Detailed Specifications*

IV. Reference Drawings and Data

On the other hand, in the case of shipyard construction, including shipyard production facilities, a pre-qualification process may not be necessary if internationally recognized construction companies are involved. However, as local construction companies usually take part, it is definitely beneficial to go through the pre-qualification process to make rational decisions. The *"Appendix B"* attached at the end of this book provides an example of pre-qualification contents for local construction companies.

6 ICT Infrastructure in the Shipyard

During the construction process of a new shipyard, it cannot be over-emphasized that the shipyard's information and communication technology (ICT) system infrastructure must be solidly built beyond the external aspect. Today, the process of establishing an ICT infrastructure should be understood as setting up an operating system for the smooth operation of a shipyard.

Depending on the size of the shipyard or the type of business, the required level of ICT infrastructure can vary significantly. The ICT system discussed here is a large-scale system for a new shipyard with an annual steel assembly capacity of 200,000 tons. As a result, it is not easy for a particular System Integration (SI) company to independently develop a successful shipyard ICT system. This is because shipbuilding experts lack the ability to accurately define the ICT system requirements that shipyards must have, and SI companies also lack knowledge of the ICT systems that shipyards need. In particular, it is important to keep in mind that relying too much on the capabilities of local SI companies will inevitably result in numerous trials and errors. Therefore, a more cautious approach is required when implementing ICT systems in the initial phases of shipyard design and construction. If there is a shipyard that can benchmark the shipyard's ICT system, such benchmarking can serve as a very solid starting point for building a new shipyard ICT system.

As mentioned later in relation to technology transfer, if a new shipyard receives support for establishing an ICT system through technology transfer from an internationally competitive shipyard, it can avoid all the trials and errors that the shipyard has experienced so far. In addition, the value of establishing an excellent operating system with competitiveness in a short period of time is immeasurable.

As shown in Figure 43, the CAD system in the early days was only applied independently in each field, such as hull-form modeling and hull structure modeling. Since then, a CAD system exclusively designed for shipbuilding, which was later expanded to include outfitting design through integrated modeling, was commercialized and began to be widely distributed among large shipyards in the mid-1990s.

Before the introduction of the ERP (Enterprise Resource Planning) system to the shipbuilding industry in the early 2000s, shipyards were computerized only in limited areas, primarily as a design-oriented ICT system. However, after the introduction of ERP, business processes were redefined according to shipyard best practices, and the scope of business computerization rapidly expanded. This made it crucial to manage a large amount of information in an integrated manner.

Furthermore, in relation to Industry 4.0, also known as the 4th Industrial Revolution, which was introduced by the chairman of the World Economic Forum in 2015, there is active promotion of the transition to this era with the concept of Shipyard 4.0 or Smart Shipyard, aiming to integrate it into the shipbuilding industry today. Moreover, ships are also transitioning to Smart Ships, often known as eco-ships, to comply with new environmental regulations. These ships are equipped with the latest information and communication technology

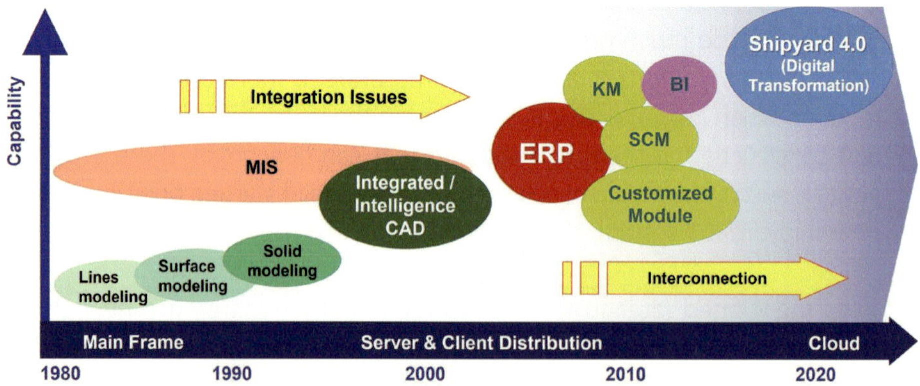

ERP (Enterprise Resource Planning) / KM (Knowledge Management) / CAD (Computer Aided Design)
MIS (Management Information System) / SCM (Supply Chain Management) / BI (Business Intelligence)
Customized modules (e.g., Planning System, Work Management, etc.)

Figure 43 Historical Trend of ICT in the Shipbuilding Industry

and are also being developed as autonomous vessels. The digital transformation of shipyards has become a crucial management innovation task today, enabling the digitization of products and services and the realization of Smart Shipyards and Smart Ships.

6.1 Information Strategy Planning

Information Strategy Planning (ISP) is significant in developing a mid- to long-term ICT master plan that considers the future direction of information and communication technology based on the business and competitive environment of the new shipyard. Generally, an ISP is established through the following steps.

Step 1: Analysis of Status

- Establishment of an information strategy direction through benchmarking ICT systems of reference shipyards
- Analyzing the internal and external environment of the shipyard
- The present and future of ICT systems and major current issues

Step 2: Determination of Strategy Direction

- ICT application principles, such as system stability, cost-effectiveness, work efficiency, and future orientation
- Establishment of an ICT plan based on informatization requirements and strategies
- Establishment of an information strategy direction and vision based on the future business environment

Step 3: Establishment of Strategy Model

- Establishment of long-term strategic goals
- Segmentation of strategic approach
- Establishment of a future ICT model
- Definition of key issues in the future ICT model

Step 4: Action Plan

- Establishment of strategic issues and tasks in the action plan
- Establishment of strategic priorities
- Establishment of an action plan
- Establishment of a verification plan through a pilot project
- Establishment of a roadmap
- Analysis of expected effects

One thing to keep in mind in the process of establishing such an ISP is that new shipyards tend to be overly eager to develop the best by simply integrating the highest standards and practices from multiple parent shipyards. First of all, it is important to note that while ICT systems can enhance the competitiveness of shipyards, they are not a universal and all-powerful solution for all problems. Unfortunately, there seem to be a few cases of shipyards attempting to build an ISP with this fantasy.

6.2 Business Architecture and ICT Architecture

Figures 44 and 45 provide exemplary illustrations of the business architecture and ICT system architecture, respectively, of a large shipyard with all functional organizations integrated.

One of the biggest dilemmas faced by new shipyards is whether to adopt an independent ICT policy and how to ensure competitiveness within a short period of time. It is not something that an SI company can provide to minimize trial and error by quickly acquiring operational expertise from the top-tier shipyards worldwide. Therefore, as mentioned above, receiving technical cooperation or technology transfer from the best partner shipyards in advanced shipbuilding countries can be the most cost-effective approach.

The essential role of ICT systems is to enable all functional organizations within a shipyard to integrate and operate on a single platform, facilitating efficient information exchange between different functions. Figure 45 is an example that illustrates the architecture of ICT systems that can achieve this objective.

Meanwhile, it is true that there are high expectations for shipyards to significantly enhance their competitiveness by implementing an ERP system. As discussed earlier in this book, there are areas within the ERP system where standard modules, such as accounting and human resources, can be easily implemented. However, due to the unique characteristics of the shipbuilding industry, certain aspects of the standard modules may need to be modified to align with the specific business processes and practices of the shipyard in question. Additionally, some modules may need to be completely newly developed based on the shipyard's unique system. In the early stages of implementing an ERP system in shipyards, it typically takes around three years to establish and standardize its optimal business processes. After the system goes live, it takes an additional three years for stabilization. Empirically, the most critical issues in introducing an ERP system include the following:

- How can business processes and the Work Breakdown Structure (WBS) be customized according to the shipyard's best practices?
- How can the shipyard's existing operating system be integrated with the standard modules of the ERP system?
- How can changes in business processes that occur after systems go-live be effectively and efficiently managed?

6.3 ICT System Construction Period and Required Professional Manpower

Even if an ICT system is established, assuming that a new shipyard to receive technological cooperation or technology transfer in the ICT field from a world-class shipyard, as shown in Figure 46, it will take at least two years. Additionally, if an ISP is established to develop an independent ICT system without sufficient benchmarking of the parent shipyard's ICT systems, it will take approximately a year to implement. As a result, the total time required to build the entire ICT system will be equivalent to the shipyard construction period.

To ensure stable operation following the implementation of an ICT system, it is necessary to secure approximately 40 expert-level operating personnel under the supervision of the Chief Information Officer (CIO), as illustrated below.

- ICT planning department: 6 people
- Operation department: 10 people
  - Network: 3 people
  - Server: 2 people
  - Database administrator: 2 people
  - Client (i.e., Computer): 3 people
- Development department: 24 people
  - Human resources, financial accounting, and cost management: 6 people
  - Material management and production planning: 6 people
  - Operation support: 12 people

Additionally, about two months of expert education and training courses are required for ICT personnel before go-live to ensure smooth operation.

- A/S (After-Sales Service)
- PR (Public Relations)
- BOM (Bill Of Materials)
- KM (Knowledge Management)
- WSD (Working Sequence Diagram)
- ICT (Information and Communication Technology)
- DAP (Detail Assembly Procedure)

Figure 44 Typical Shipbuilding Business Architecture

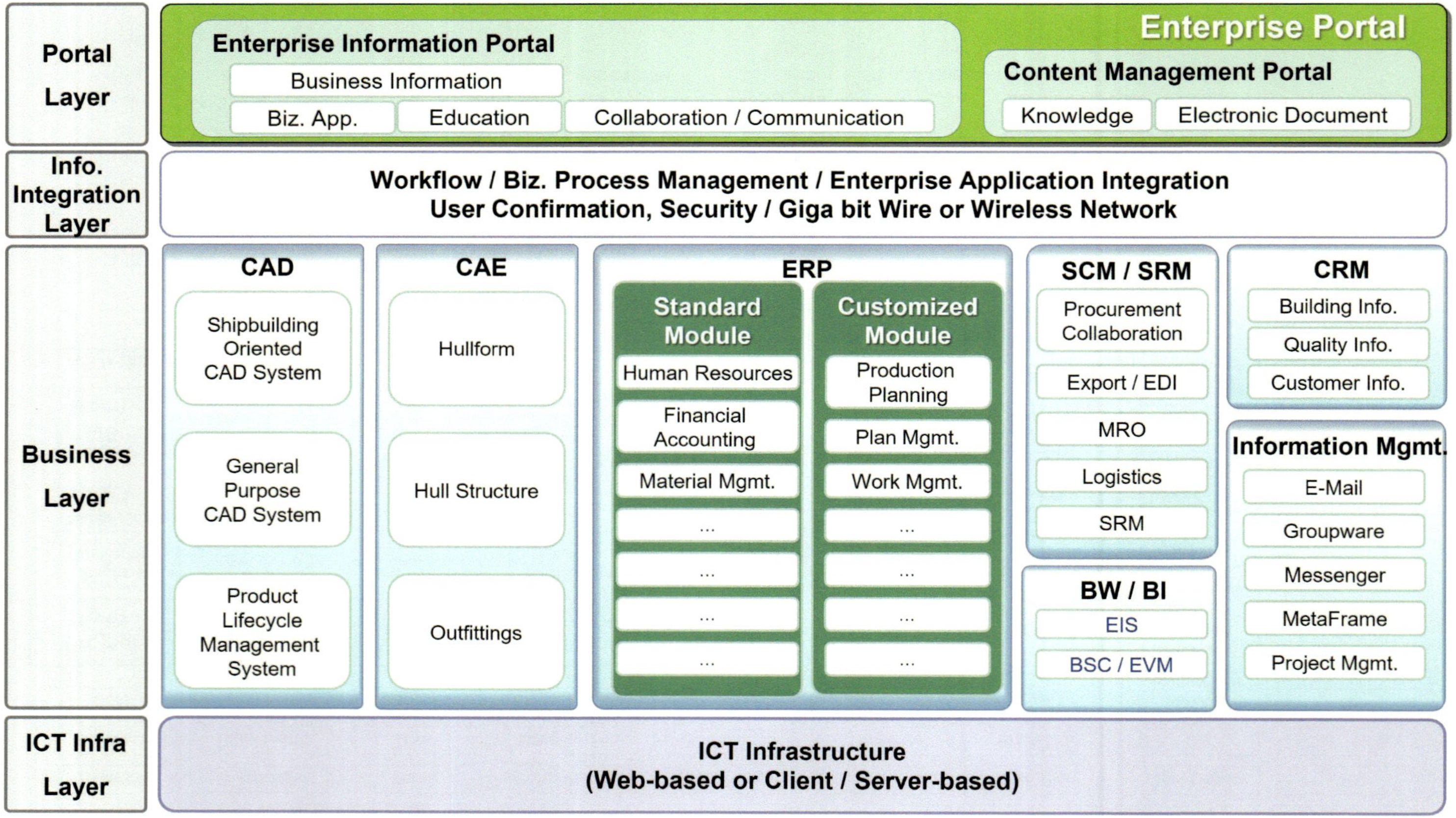

- BI (Business Intelligence)
- CAD (Computer Aided Design)
- EDI (Electronic Data Interface)
- EVM (Earning Value Management)
- SRM (Supplier Relationship Management)
- BSC (Balanced Score Card)
- CAE (Computer Aided Engineering)
- EIS (Executive Information System)
- MRO (Maintenance, Repair & Operation)
- BW (Business Warehouse)
- CRM (Customer Relationship Management)
- ERP (Enterprise Resource Planning)
- SCM (Supply Chain Management)

Figure 45 Typical Shipyard ICT Systems Architecture

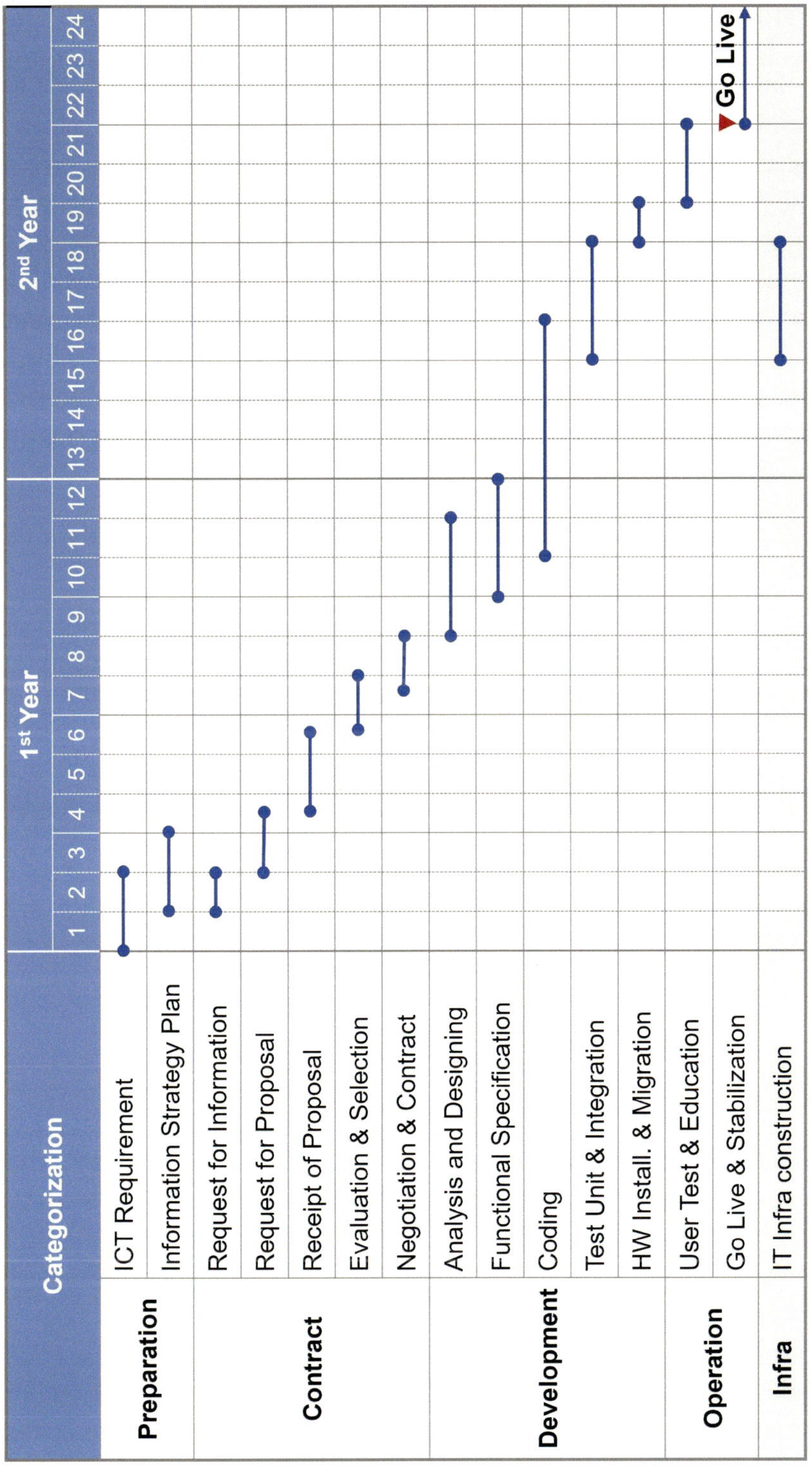

** Remark: ICT Data Center & Network construction should be completed before user test.*

Figure 46 Typical Timeframe of ICT Systems Implementation

6.4 Shipyard 4.0

In terms of the future development direction of the shipbuilding industry, the transition to eco-friendly smart ships and smart shipyards (also known as Shipyard 4.0) represents an unprecedented task for the future. Of course, these changes can pose significant challenges in political, economic, and social aspects, but they can also serve as a catalyst for competitiveness for shipyards in the future. Shipyard 4.0 can be simply defined as the integration of the 4th Industrial Revolution into shipyards through digital transformation. According to Wikipedia, the primary objective of the 4th Industrial Revolution is to enhance operational efficiency. The key factors driving this revolution are outlined below.

- Digitization and integration of vertical and horizontal value chains
- Digitization of products and services
- Digital business models and customer access

In addition, there are four key technologies that lead this digital transformation.

- Cyber-physical systems (e.g., digital twin)
- Internet of Things (IoT) and smart sensors
- Cloud computing
- Artificial intelligence computing

Figure 47 depicts a conceptual schematic diagram illustrating the objectives of Shipyard 4.0 and Smart Ship. The core of this concept involves integrating the core technologies of the 4th Industrial Revolution into the shipyard operation process and ship operation elements. In general, when new management theories, techniques, and technologies emerge, everyone tends to be enthusiastic and actively embrace the latest trends. However, the latest core technologies used in the Fourth Industrial Revolution are only tools and not the final goal itself. Therefore, when applying this concept, it is important to be careful not to confuse or distort the final goal and the tools.

The objectives of Shipyard 4.0 may vary depending on each shipyard's management strategy, but it is desirable to aim to achieve the following general goals to enhance operational efficiency as a smart shipyard.

Lead Time Innovation

- Reduce the construction period (including reducing the inspection scope for shipowners and classification society surveyors).
- Issue design drawings and supply materials on time.
- Minimize buffer time and rush tasks.
- Enhance Total Productive Maintenance (TPM) activities, etc.

Productivity Innovation

- Reduce workload.
- Reduce design and production man-hours.
- Maintain and improve progress compliance rates.
- Expand automated production processes, etc.

Zero Loss

- Eliminate quality failure costs.
- Eliminate waiting times for logistics and waste.
- Eliminate loss and damage to materials.
- Minimize surplus materials, etc.

The introduction of Shipyard 4.0 is not a task that can be achieved in a short time, so it is desirable to promote it in line with the development trend of Smart Ship with mid- and long-term plans. Rather than this, it is necessary to first focus more on establishing a strong foundation for the fundamental operational processes of the shipyard itself.

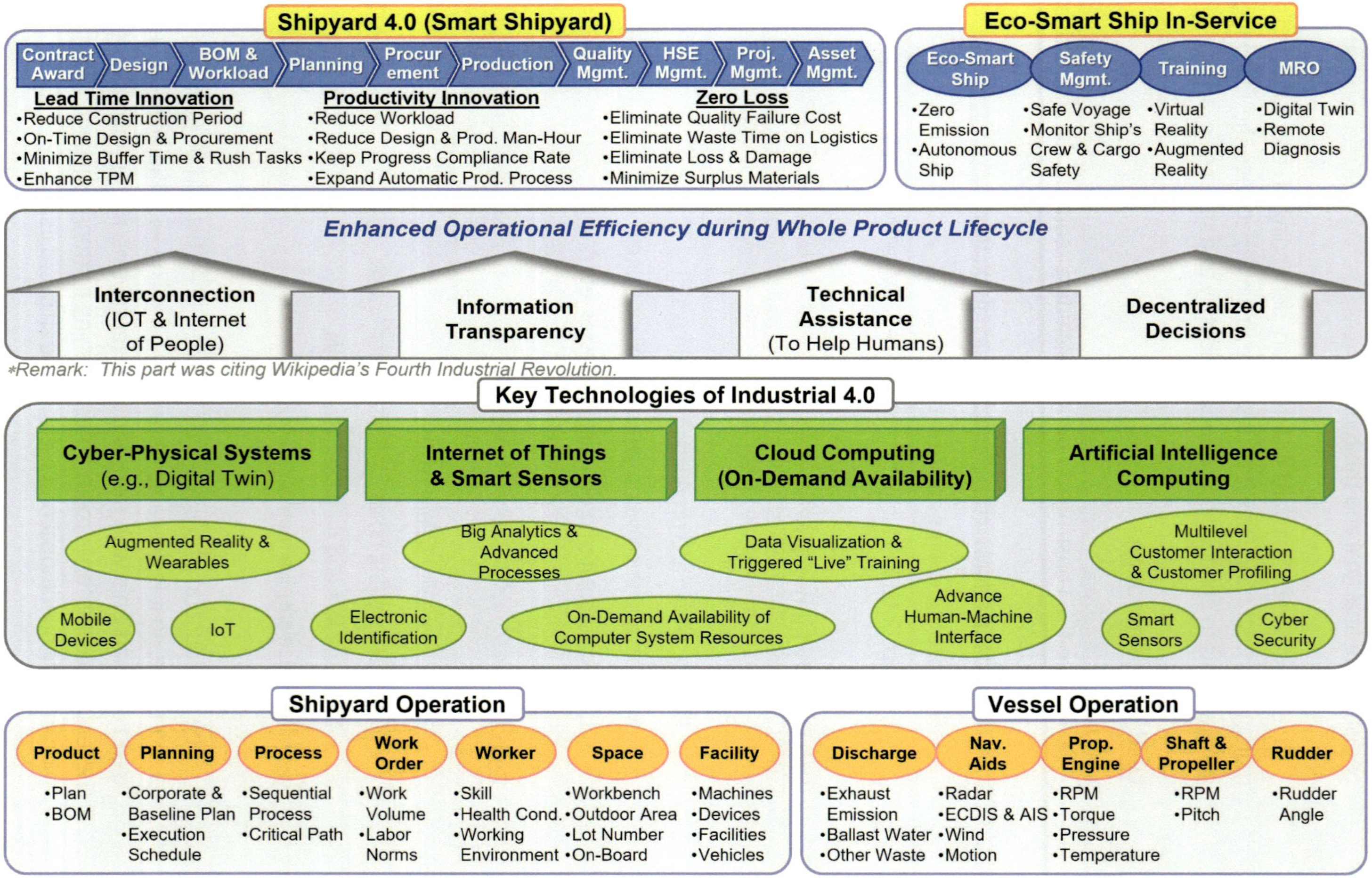

Figure 47 Conceptual Diagram of Shipyard 4.0

7 Training and Technology Transfer

Along with establishing ICT systems, technology transfer, and employee education and training are crucial components from a software perspective in shipyard operations. Furthermore, setting up an operating system for a new shipyard independently demands considerable time and effort. Even with the recruitment of a few talented individuals with extensive experience in advanced shipyards, there are numerous limitations and trial-and-error processes involved in constructing a reliable operating system. Therefore, it should be acknowledged that systematically acquiring shipyard operation expertise from advanced shipyards is the most effective and efficient way to quickly secure competitiveness and technology in the most economical manner.

7.1 Considerations of Training and Technology Transfer

Technological Solution Partner

Every leading shipyard around the world has its own unique characteristics. Even for ships of the same type, each shipyard has its own unique practices regarding design details and various operations related to production planning and production. As a practical example, one of the mistakes that must be avoided is attempting to easily establish best practices for a new shipyard by solely applying ICT technology to the individual strengths of several shipyards without a comprehensive understanding of the technical background. This simplistic approach can only lead to considerable confusion because of its complexity.

Therefore, considering the relationships outlined below, it is desirable to select the most appropriate technology transfer partner from the collaborative relationships between advanced shipyards possessing technology and latecomer shipyards in need of technology.

- There is little conflict of interest in business areas or products between each other.
- Mutually complementary relationships in business areas are essential for success.

- A win-win relationship can be established outside the shipbuilding industry through offset trade, among other methods.

Generally, it may not be realistic to use the operating system of an advanced shipyard as it is influenced by numerous factors such as regional lifestyle, working mindset, standard of living, and climatic conditions. Therefore, it is a more reasonable approach to gradually localize after receiving the technology transfer and develop it into the new shipyard's own technologies.

Training and Technology Transfer Fields

First of all, when setting up a shipyard operating system, it is crucial to recruit top-tier personnel from diverse fields. This includes engineers responsible for design, workers involved in production, and managers overseeing key functional departments in shipyard operations. In conjunction with the construction of a new shipyard, it would be more practical to send a carefully selected group of elite personnel to an advanced shipyard for intensive education and training. Upon their return, they can then provide training to a larger group of employees in the field. Similarly, design, production, and management should be centralized at a single advanced shipyard, rather than being dispersed across various shipyards for education and training.

In addition to training on the type of ship that will be the main focus of the shipyard in the future, specialized education and training should also be implemented for specific ship types or products. This includes cargo containment construction for LNGC cargo holds, offshore topside module assembly, and more.

Training Stage and Place

If the local vocational training and education system in the country is not well-equipped, it is necessary to establish a training center at the shipyard. Having a training facility available during the shipyard construction phase is highly beneficial for expediting the commencement of shipbuilding at the new shipyard. Therefore, it is recommended to plan education and training in at least four stages as follows:

- Overseas dispatch training for technology transfer
  - Key personnel selection: This includes managers, engineers, and foremen at the lowest level of each organizational structure, as well as instructors at the local training center.
  - Education and training content: The content comprises theoretical education as well as practical training, which may also include on-the-job training if necessary.
  - Training location: Advanced shipyard for technology transfer.
- Dissemination training at the shipyard construction stage
  - Local training is provided by the shipyard's instructors for both existing and new employees.
  - Education and training content: Theoretical education and practical training.
  - Training location: Local training center.
- Technical advisor guidance
  - Technical advisors from an advanced shipyard temporarily reside in new shipyards and provide technical guidance until a certain level of technological proficiency is achieved.
  - Troubleshooting and providing technical guidance for various practical issues that arise at the local shipyard.
- New employee training in response to increased production volume
  - Education and training are provided to newly hired personnel based on annual manpower requirements.
  - Training location: Local training center.

Local Instructors and Trainees

In order to minimize the cost of education and training for overseas assignments, it is preferable to select trainees who already possess a certain level of knowledge and skills, rather than hiring new employees. When implementing technology transfer in advanced shipyards, it is necessary to issue a certificate confirming that the trainees have reached a level where they can disseminate the knowledge acquired during their education and training upon their return to their home shipyard.

Unlike in Korea, overseas hull welders generally refer to welders with excellent skills who are capable of welding in all positions. However, Korean shipyards employ a Build Method that ensures the stability of welding quality by enabling welders to work in the most comfortable position. Therefore, a hull welder meets the minimum qualifications if they hold a welding license for at least 3G and 3F. In other words, even if a new production worker has no experience in welding, they can obtain a hull welding license with about four months of training in welding at the shipyard training center.

Training Materials

Various types of textbooks are utilized in education and training processes, but recently, with the advancement of ICT, education through media platforms like YouTube, virtual reality, and augmented reality has become feasible. Nevertheless, we must not forget that the active process of thinking and acting on one's work is more important than simply observing and listening.

On the other hand, intellectual property is crucial in technology transfer, and it is absolutely necessary to have a mutual agreement on the right to use, as well as the provision of related information. In addition to drawings of ships, the representative intellectual properties of shipyards include designs and quality standards, manuals, guidelines, practices, and safe working procedures, among others.

7.2 Overall Training Scheme

In order to establish a comprehensive education and training scheme, several prerequisites must be determined in advance. These include:

- Deciding whether to produce the steel structure required for the construction of the shipyard workshop in-house.
- Determining whether the dock gate, Goliath crane structures, floating dock, barge, and other equipment required for shipyard construction

will be built by the shipyard itself as a pilot project for shipyard operation.

- Deciding whether to place an order a leading ship for On-the-Job Training (OJT) of employees to be dispatched to an advanced shipyard for technology transfer, or to construct a ship through joint collaboration between the two shipyards.

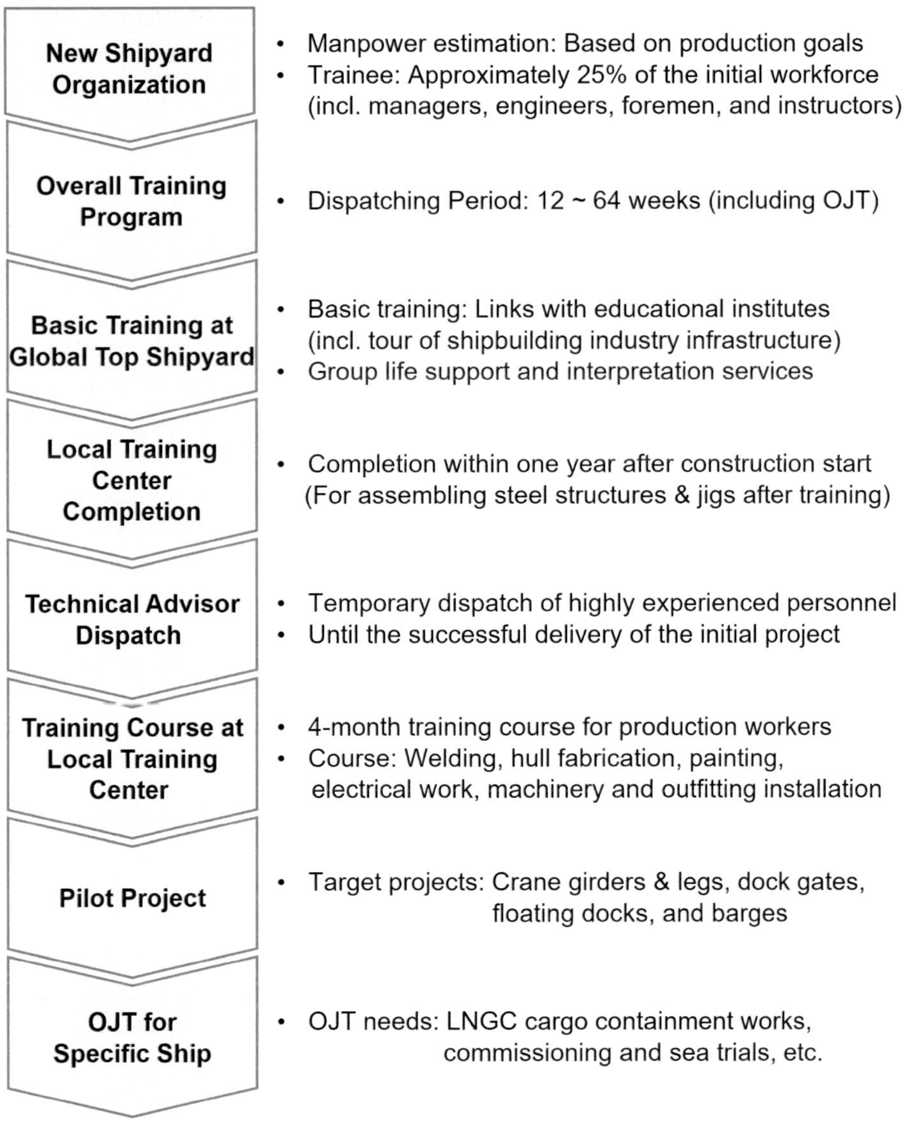

Figure 48 Considerations for Establishing an Overall Training Scheme

Once the preconditions for establishing education and training plans have been determined, the overall training scheme can be finalized by referring to the steps and considerations outlined in Figure 48.

7.3 Phase of Technology Transfer

As an ideal approach to transfer shipbuilding technology from the global top-tier shipyards, it is most desirable to implement it in the three phases described below. The overall concept is depicted in Figure 49.

- Leading ship construction at an advanced shipyard
- Joint construction at both shipyards
- Self-construction at a local shipyard

In general, technical cooperation is not easily achieved by simply requesting technology transfer from a leading global shipyard. This is because such collaboration could potentially cultivate future competitors. Therefore, it should be noted that the probability of technology transfer can be increased only if there are attractive conditions associated with it. As part of this approach, it is common to place an order for ships to be built at a globally competitive advanced shipyard and to include contract provisions related to technology transfer.

If two shipyards are located relatively close together, a Joint Construction phase between them may be considered. Since the construction process for the stern part of a ship is relatively complex, it is a construction method in which the ship is first built in the form of a half-ship at an advanced shipyard and then transferred to a local shipyard to finally complete the entire ship. This collaboration method has the advantage of being able to conduct OJT along with education and training courses at advanced shipyards during the shipbuilding process. This is an effective method that can be applied in cases where a relatively high level of shipbuilding technology is required, such as LNGC and offshore products. However, one of the limitations of this method is that if the distance between the two shipyards is too far, excessive costs due to dry towing of the half-ship are incurred.

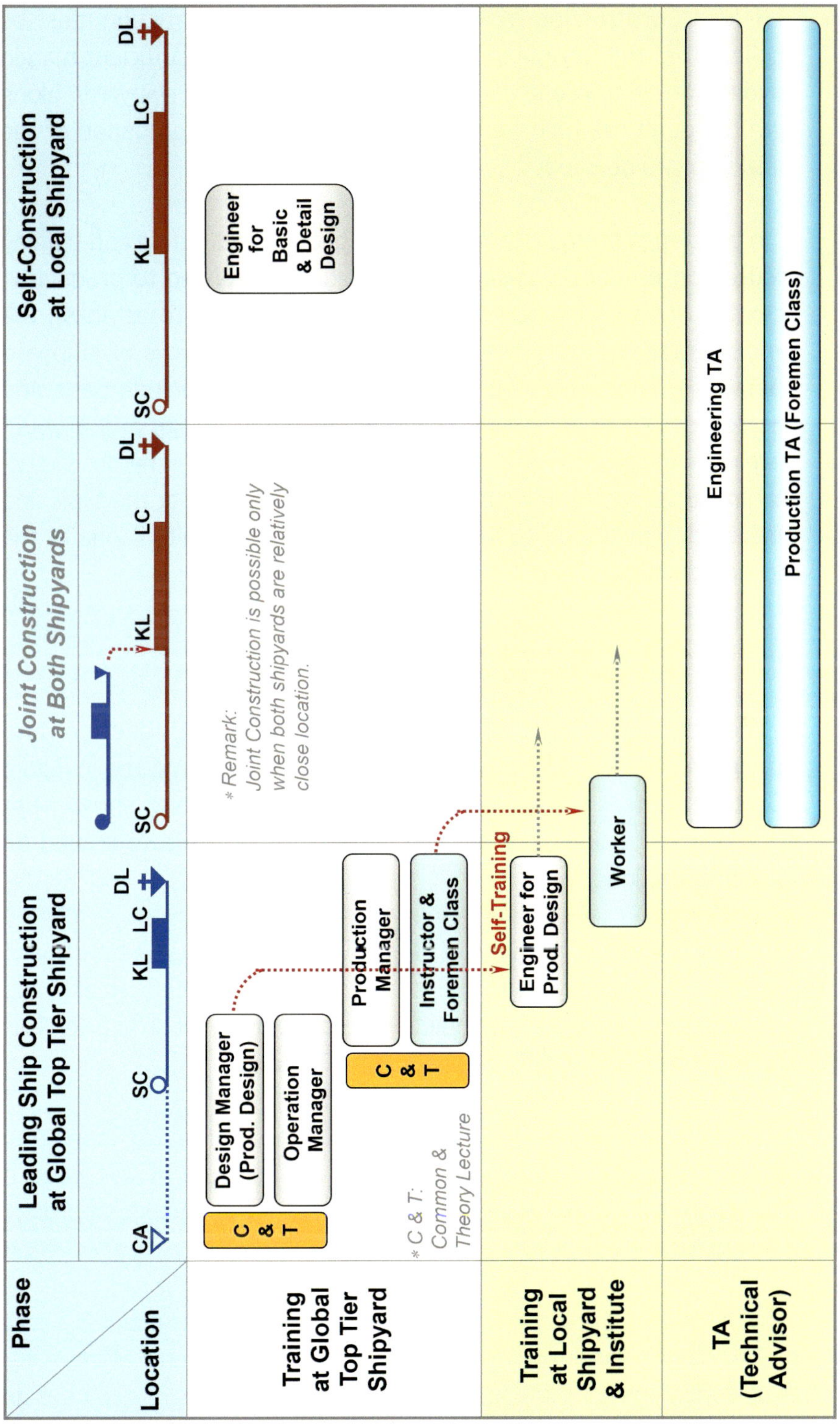

Figure 49 Typical Concept of Training and Technology Transfer Phase

If Joint Construction is not possible and Self-Construction must be applied from the beginning, it is necessary to have a sufficient margin beyond the difference in productivity between the two shipyards, along with thorough risk management during the construction period of the initial shipyard operations.

In the design sector, it is important to prioritize education and training focused on production design, which is directly related to production activities. This is because it is absolutely necessary to establish the fundamental framework that serves as the central axis of shipyard operations. Then, the scope must be expanded to include basic and detailed design in order to develop independent design and product development capabilities. In theory, the majority of the numerous factors that influence the overall cost of a ship are determined at the basic and detailed design stages, so independence of design ability is one of the most crucial factors.

7.4 Duration of Training for Technology Transfer

The education and training period for design, production, and management fields related to technology transfer is illustrated in Figure 50. It references shipyard operational experience and shows that education and training courses can range from 12 weeks to around 64 weeks in duration.

All personnel in the shipyard should have a basic understanding of common education, which includes subjects covering Safety, Health, and Environment (SHE), the shipbuilding process, and hull structure. This education typically lasts for approximately 6 weeks.

Shipyard managers acquire the basic capabilities of managing new shipyards by completing an additional 10 weeks of OJT in each management field, following their collective training in general project management.

In the case of the shipyard design organization, it takes about 8 weeks of CAD training for design drawing, in addition to common training. After that, depending on the specific working characteristics of each design

Course		Total Week	Common & Basic Theory			Practical Theory and Training			OJT
Training Organization			Institute or Training Center			Training Center of Global Top Tier Shipyard			Each Field
Duration			1 Week	1.5 Week	1.5 Week	2 Week	8 Weeks	12 / 48 Weeks	8 ~ 40 Weeks
Management		16	Industry Safety, Health & Environment & Shipbuilding Industry	Shipbuilding Process	Ship Structure	PM (General)	-	-	10
Design		40~52				-	Shipbuilding Drawing & CAD	-	28~40
Production (Foreman & Key Worker)	Fabrication	28				-	-	12	12
	Assembly								
	Welding								
	Painting								
	Outfitting Fabrication								
	Outfitting Installation								
	Machinery Installation								
	Electric Outfitting								
	Test and Trial	40						12	24
	LNGC Production	*64*						*Invar Welding : 48*	*12*
	Joinery Work	12						-	8
	Production Support	12							
	Production								
	Production Logistic								
	Dock Operation								
	Maintenance	12							
	C & A								
	Facility								
	Machine								
	Heavy EQPT. Supporting								

* Remark: Outfitting (means Pipe, Duct & Hull Outfit), C&A (Civil & Architect)

Figure 50 Typical Duration of Training for Technology Transfer

field, an additional 28 to 40 weeks of OJT is conducted in each design department.

In the production field, there is a significant difference in the training duration between direct and indirect production organizations. In the case of an indirect production department, employees can further develop their skills by undergoing OJT for approximately 8 weeks after completing the common training program. Except for cases requiring special skills such as the commissioning process or LNGC cargo containment process, trainees undergo approximately 12 weeks of education and training for direct production areas such as steel fabrication, assembly, welding, painting, pre- and post-outfitting, machinery installation, and electrical works before obtaining relevant licenses. After that, they are assigned to each production department and undergo an additional 12 weeks of OJT to further develop the necessary skills. However, when it comes to commissioning tasks, which are the final processes of shipbuilding, it is necessary to accumulate approximately 24 weeks of experience through OJT. In the case of LNGC cargo containment tasks, it takes about a year to acquire the specialized welding skills required for cargo hold construction.

7.5 Required Manpower vs. Number of Trainees for Technology Transfer

The total workforce needed for a new shipyard must be calculated by carefully analyzing numerous parameters, including the primary type of ship to be built, construction period, mid- and long-term production program, productivity, and WBS system. In addition, although there may be variations depending on the size or organizational structure of the shipyard, it is necessary to provide direct technology transfer training from a leading global shipyard to approximately 25% of the total workforce required during the initial phase of shipyard operation. These key personnel need to return home after completing relevant training and provide local dissemination training to newly recruited personnel. Therefore, it is necessary to verify fundamental skills and loyalty to the company in advance.

Table 35 below illustrates an example of calculating the required workforce based on the trend of increasing production volume and improving productivity in a shipyard with an assembly production capacity of up to 200,000 tons per year. This table also compares the workforce composition ratio of a leading global shipyard, assuming a functional organizational structure. Based on this, it is possible to determine the rational demand for dispatching personnel to receive direct technology transfer from a leading global shipyard at the beginning of shipyard construction. Since this part is highly variable and dependent on the project's workload, it is preferable to view it as a methodology rather than solely focusing on the end result figure.

Organization	Trainee	Local Shipyard Employees			Global Shipyard
		Stage 1	Stage 2	Stage 3	
Total (Persons)	**666**	**2,500**	**5,232**	**7,111**	**100%**
Management Division	**58**	**101**	**208**	**275**	**3.8%**
Marketing	2	4	7	15	0.3%
Quality Management	41	71	150	188	2.7%
Health, Safety & Environment	14	23	46	61	0.8%
Legal Affairs Staff	1	3	5	11	0.04%
Corporate Governance	**27**	**66**	**126**	**140**	**1.9%**
Human Resources	4	12	25	31	0.4%
General Affairs & Admin.	20	45	84	89	1.4%
Corporate Management	3	9	17	20	0.1%
Engineering and Technology	**80**	**176**	**290**	**623**	**9.0%**
Sales Engineering	4	7	7	20	0.7%
Integrated Design and Design Planning	4	8	14	31	0.5%
Hull Design	8	19	34	91	1.2%
Machinery & Outfitting Design	25	62	112	279	3.8%
Electrical Design	10	25	45	100	1.5%
ICT Planning	12	31	42	42	0.5%
Industrial R&D	4	11	11	29	0.4%
Procurement	13	13	25	31	0.4%

Organization	Trainee	Local Shipyard Employees			Global Shipyard
		Stage 1	Stage 2	Stage 3	
Production	**495**	**2,141**	**4,592**	**6,055**	**85.1%**
Project Management	7	19	26	50	0.9%
Production Management	22	34	63	81	1.1%
Facility Maintenance	25	62	93	120	2.2%
Subcontract Management	3	4	10	21	0.3%
Indoor Production	298	1,172	2,051	2,205	23.4%
Outdoor Production	99	683	2,042	3,080	46.0%
Production Support	41	167	307	499	11.2%

Table 35 Typical Number of Trainees for Technology Transfer

7.6 Training Program at the Vocational Training Center

Here are the education and training subjects offered in the affiliated vocational training centers of the leading global shipyard, along with their contents. The training period for all subjects typically lasts for a 4-month course.

Generally, there are no specific qualifications required for admission to vocational training centers, except for age restrictions.

Ship Design Training Course

Training Contents	Major Curriculum
• Hull Design • Piping Design • Hull Outfitting Design • Accommodation Design • Machinery Design • Electrical Design	• Safety • Introduction to Ship Structures • Introduction to Each Design Discipline & Drawing (Hull, Outfitting, Accommodation, Machinery, & Electrical Design) • Criteria and Guidelines for Production Drawings • Study of Standards and Practices • Practice of CAD

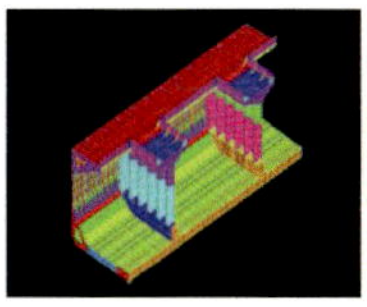
[Hull Design]

[Outfitting Design]

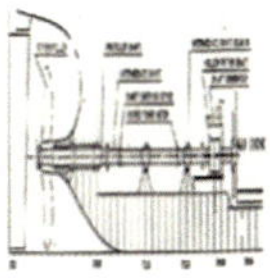
[Machinery Design]

[Accommodation Design]

[Electrical Design]

Hull Fabrication Training Course

Training Contents	Major Curriculum
• Theory and Practice • Structure Fabrication • Line Heating • Cutting - Plate NC Cutting - Manual Cutting • CO_2 Welding • Handling of Measuring Tools	• Safety - General - Safety of Gas Handling and Shipbuilding Materials - Safety of Welding • Understanding Production Drawing • Fabrication Process and Criteria • Fabrication Method • Characteristics of Gas, Handling of Cutting Tools, and Cutting Methods • Accuracy Measurement and Tool Handling • Study of Welding Materials • Basic and Practice of CO_2 Welding • Removal Method for Attachments • Equipment Maintenance

[Structure Fabrication]

[Line Heating]

[NC & Manual Cutting]

[Steel Piece Removal]

[Measuring Device]

[CO_2 Fillet Welding]

Welding Training Course

Training Contents	Major Curriculum
• Theory and Practice • CO_2 Welding • TIG Welding • Electro Gas ARC Welding (EGW) • Submerged ARC Welding (SAW)	• Safety - General - Safety of Gas Handling and Shipbuilding Materials - Safety of Welding • Basic and Practice of CO_2, TIG, EGW, and SAW Welding • Removal Method for Steel Pieces • Defect Remedy • Equipment Maintenance

[CO_2 Welding]

[TIG Welding]

[EGW Welding]

[SAW Welding]

Machinery and Outfitting Installation Training Course

Training Contents	Major Curriculum
• Theory and Practice • Installation of Steel Outfitting • Installation of Instrumentation • Installation of Machinery • Machining • Commissioning & Trial	• Safety - General - Safety of Gas Handling & Shipbuilding Materials - Safety of Welding • Understanding Pre-Outfitting, Machinery, CO_2 Welding, Gas Cutting, and Outfitting Quality • Introduction and Installation of Steel Outfitting, Pipes, and Valves • Introduction and Installation of Instrumentation • Introduction and Installation of Machinery & Auxiliary (Main Engine, Shaft, Rudder, Aux. Machinery, etc.) • Introduction and Operation of Machine Tools, Alignment Methods, and Assembling Machines • Machining • Commissioning • Introduction and Practice of Sea Trial

[Outfittings Installation]

[Instrument Installation]

[Machinery Installation]

[Machining]

[Commissioning & Trial]

Painting Training Course

Training Contents	Major Curriculum
• Theory and Practice • Blasting and Grinding • Shipbuilding Paint • Painting General and Color Study	• Safety - General - Safety of Gas Handling and Shipbuilding Materials - Safety of Welding • Study of Hull Structures • General for Blasting and Painting • Grinding • Touch-up and Spray Works • Facility for Airless Painting • Painting Quality and Color Scheme

[Blasting & Grinding]

[Airless Painting]

[Touch-Up Painting]

[Painting & Color Scheme]

Electrical Works Training Course

Training Contents	Major Curriculum
• Theory and Practice for Electrical and Electronic Works • Cable Laying and Connection • Installation of Electrical Equipment (Control Panels, Lighting, etc.) • Instrumentation Works • Commissioning of Electrical Equipment	• Safety - General - Safety of Gas Handling and Shipbuilding Materials - Safety of Welding • Introduction to various types of Electrical Equipment, including Panels, Motors, Generators, Transformers, Lighting and Sensors • Cable Laying and Connection • Installation of all types of Electrical Equipment • Installation and Adjustment of Instrumentation • Installation of Navigation and Communication System • Technical Manual Study • Commissioning • Introduction and Practice of Sea Trial

[Electrical EQPT Installation]

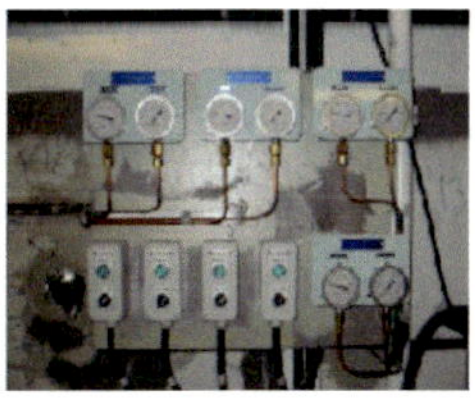

[Instrumentation Works]

[Cable Laying & Connection]

A Place
Where
Ships are Built

Part III.

Valuation for a New Shipyard Operation

This is an example of how shipbuilding lead time can be shortened and sales increased by outsourcing the production of a super-large Ring type block, transporting it to the shipyard using a heavy cargo carrier, and then erecting it into a floating dock with the assistance of a 3600-ton floating crane.

Part III.

Valuation for a New Shipyard Operation

In the 2000s, numerous countries with abundant resources have implemented policies to establish new shipyards or attract Foreign Direct Investment (FDI) from international companies in order to foster their own shipbuilding industries. As previously mentioned, the shipbuilding industry not only generates significant employment opportunities, but the establishment of new shipyards also has a positive ripple effect on other industries. However, due to the substantial investment in this industry, it is unrealistic to expect profits to be generated quickly enough to recoup the investment in a short period of time. Therefore, in some countries, the government bears the initial Capital Expenditure (CAPEX) needed to construct a shipyard.

Therefore, even though there are many variables, it is necessary to verify investment risks through a corporate valuation simulation based on a business plan for at least 10 years after the shipyard's construction. Furthermore, it is important to check whether additional government support policies are needed before making a final investment decision.

There are various methods of enterprise valuation, but the Discounted Cash Flow (DCF) model assesses the present value of the FCF (Free Cash Flow) by subtracting the investment amount from the cash flow generated through the new shipyard's operating activities. Here, the factors to consider from the perspective of shipyard operations are described, among the various variables that arise in this process.

1 Valuation Assumptions

In order to reasonably evaluate the value of a new shipyard, it is essential that key conditions related to shipyard operation have universal applicability.

Production Program Based on Long-Term Operation

For the valuation of a new shipyard, a production program should first be prepared. This production program should systematically and logically define all major production events for a potential project over approximately 10 years during the initial phase of the shipyard's operations. It can be written by referring to the production program provided as a sample in Figure 13 in Part II, Chapter 2. Additionally, this production program should, in principle, include the following information:

- Type of ship and number of ships to be built per year
- Key event milestones for each project (i.e., steel cutting date, keel laying date, launching date, and delivery date)
- Shipyard construction schedule and shipyard partial operation schedule
- Productivity improvement plan (i.e., construction period and number of production man-hours required by ship type)
- Load balancing of resources (i.e., annual production capacity and human resources)
- Available man-hours per year, etc.

While it may be nearly impossible to create a production program and assess a corporate valuation during the initial decision-making stage, it is important to view it as a necessary step before commencing actual construction work. Accordingly, the participation of shipbuilding planning experts, who can establish mid- and long-term production plans, is essential to carry out this complicated work.

Capital Expenditures (CAPEX) Estimation

Although the standard for the useful life of tangible assets may vary slightly from country to country, it is convenient to categorize and aggregate the initial investment cost estimate as shown in Table 36 by referring to the depreciation useful life below.

Additionally, when calculating the initial investment cost, all investment costs for each stage of shipyard development, as outlined in the mid- and long-term plan, must be included. When estimating investment costs using reference values for similar construction projects, it is important to carefully consider significant variations in unit costs based on the type of construction and the country. Therefore, it is necessary to allocate a certain amount of contingency budget to account for other unknown variables.

Categorization		Useful Life	Investment Amount
Civil	Dry Dock, Quay, Crane Rail, Dredging, Earthworks, Utilities, etc.	50 years	
	Road and Pavement	12 years	
Shop & Building	Workshop, Office, Substations, Compressor Room, etc.	25~50 years	
Structure	Dock Gate, Mooring Facilities, Gas & Water Tanks, Lighting Tower, etc.	15~50 years	
Equipment	Crane, Welding & Cutting Machine, Painting Equipment, Lathe, etc.	12 years	
Vehicle & Ship	Transporter, Truck, Mobile Crane, Cherry Picker, Table Lift, Forklift, etc.	6 years	
	Tugboat, Floating Crane, Aircraft, etc.	15 years	
Tool & Jig	Workbench, Stool, Pin Jig, Trestle, Sailing Jig, Rope, etc.	6 years	
Other Facilities	Office Facilities, Test and Measuring Devices, Production Aids, Cooling and Heating Facilities, etc.	6 years	
ICT System	Hardware, Software, SI Fee, etc.	6 years	
Total Amount of Investment	Net Tangible Assets	-	
	Contingency (%)		

Table 36 Depreciation Useful Life

2 Simulation Logic

2.1 Revenue

Ship Price Estimation

When estimating the ship price, if the current ship price has already been agreed upon, the contracted ship price can be applied. Additionally, future ship prices can be estimated using past price trends by ship type provided by Clarksons Research, a globally recognized information organization. Based on these past trends in ship prices, the Cumulative Annual Growth Rate (CAGR) can be calculated and then effectively used to estimate ship prices for specific years in the future.

For reference, there may be variables depending on the type of ship and the timing of data extraction, but the annual rate of new shipbuilding price growth shows an approximate range of 2.5% ~ 3.1% statistically.

Ways of Sales Recognition

There are differences in the construction duration for each type of ship, and depending on the characteristics of the ship, the cost-input ratio can vary significantly at various stages of the project. Additionally, due to the significant variation in the cost-input ratio depending on the build method, it is impossible to define a standard representative value. Therefore, consistency and rationality can be ensured by deriving and utilizing cost-input ratio information that can be referenced in similar projects.

Recognition of sales during the shipbuilding period can be simplified as follows.

- The formula to calculate the amount of sales by shipbuilding stage

  = Ship Price x Cost-Input Ratio by Shipbuilding Stage

Here,

- Cost-Input Ratio = Actual Input Cost / Total Planned or Target Cost in the Business Plan

- Actual Input Cost: The cost includes both the direct costs required for the construction of the project and the indirect distribution costs incurred to support production activities.

- Total Planned Cost: The estimated total cost or target total cost applied to the price quotation in the business process.

In valuation, it is most reasonable to calculate the cost-input ratio corresponding to the key events schedule of similar projects. Therefore, based on the reference values in Table 37 below, the average value is uniformly applied to each period between production key events, or the S-Curve is derived from similar projects to calculate sales by project period.

In the case of a ship type for which there is no similar project data, one possible method is to calculate using interpolation or extrapolation based on the total construction period and the reference values provided in Table 37.

Categorization	SC - 3M	Steel Cutting	Keel Laying	Launching	Delivery
COT	1.5%	6.5%	77.2%	97.1%	100%
LNGC	0.3%	1.3%	45.8%	73.0%	100%

Table 37 Typical Accumulated Ratio of Actual Input Cost by Key Events

2.2 Cost of Goods Sold

The following is an overview of the matters to be considered for each major account of COGS (Cost of Goods Sold).

Material Cost

Among the costs of sales, the recognition of material costs during the construction period of the ship is calculated in the same manner as the recognition of sales by period, based on the cost-input ratio.

- Typical estimation of material costs by year
  - Proportion of material cost: COT ≈ 60±α% of the ship price
    LNGC ≈ 50±α% of the ship price
  - Material cost escalation by year: Apply the Compound Annual Growth Rate (CAGR) value when estimating future ship prices.

Table 38 below provides a reference value of the material cost-to-input ratio by ship type.

Categorization	SC - 2M	Steel Cutting	Keel Laying	Launching	Delivery
COT	0.1%	1.9%	79.3%	96.7%	100%
LNGC	0.01%	0.1%	48.5%	77.3%	100%

Table 38 Typical Accumulated Ratio of Material Cost by Key Events

Labor Cost

Prior to estimating labor costs, the productivity improvement target index, which reflects the learning effect year by year, should be calculated. Once this index is calculated, the production program for each project can be determined during the construction period, taking into account the index. Based on this production program, the annual assembly volume (or weight) can be determined, and the man-hour job requirement can be calculated. These calculations enable the determination of the annual manpower requirements and the minimum manpower needed to handle the lowest production levels on a monthly basis.

The production man-hours by ship type, which are used as the standard for calculating the required manpower, are trade secrets for each shipyard and are not specifically mentioned here. If the minimum required manpower for each year is estimated, the labor cost can be calculated by referencing the wage level in the country based on the level of experience or education.

Meanwhile, in order to establish the standard for calculating labor costs in the country, the necessary experience for each level of shipyard personnel is determined by referring to the sample provided in Table 39 below.

Categorization	Position	Career Level	Salaries
Management	Managing Director	CEO	
	Deputy MD	Vice president	
	Chief of Division	Executive VP	
	Chief of Team	Director	
Office Worker (University Graduate)	Chief of Group	Over 19-year of career	
	Senior Manager	Over 19-year of career	
	Junior Manager	Over 13-year of career	
	Manager	Over 8-year of career	
	Assistant Manager	Over 3-year of career	
	Freshman	Freshman	
Production Worker	Master	Over 20-year of career	
	Foreman	Over 15-year of career	
	Worker	Over 10-year of career	
	Freshman	Freshman	
	Outsourced worker	Outsourced worker	

Table 39 Typical Career Requirements of Manpower Hierarchy in the Shipyard

In the case of a foreman who directly leads the workers at the production site, it is generally appropriate to consider one out of every 15 to 20 workers. Since it is not always possible to maintain a consistent workload throughout the construction period, the minimum required manpower is calculated based on the lowest monthly workload of the year. It is desirable to ensure that overtime work is completed within the legal limits of the country, but if additional manpower shortages occur, mobilizing outsourced manpower may be a viable solution.

When calculating annual labor costs, one approach to incorporate the wage increase rate is to utilize historical statistics over an extended period of time. Alternatively, if relevant data is unavailable, one can use inflation rate statistics from the past 10 years. With reference to these matters, labor costs are calculated for each year using the formula below.

- Labor Cost = Σ(Number of Employees by Workforce Hierarchy x Applicable Salaries including Bonus) + Σ (Sum of Overtime by Employee by Workforce Hierarchy x Applicable Overtime Wage)

If there are any outsourced labor costs, as mentioned later, they are recorded in a separate account to differentiate them from direct labor costs.

Training and Technical Advisor Fee

Education and training costs are not specifically addressed here, but if they are considered as a certain percentage of labor costs, they can be allocated based on the number of employees added each year. When considering training dispatched to a leading global shipyard, the additional costs incurred will be significant and should never be overlooked. Furthermore, if technical advisors are dispatched from a leading global shipyard and remain on site until the construction quality of the new shipyard is established, substantial costs will also be accrued. Therefore, efforts must be made to minimize the extent and duration of their involvement.

Engineering Package Service Fee

As it takes a considerable amount of time for new shipyards to develop their own engineering capabilities, it is common practice for them to receive all engineering packages, including design drawings for basic design, detailed design, and production design, from a leading global shipyard or an engineering company at the initial stage. At this time, the engineering package service cost, including royalties, is incurred before the actual production period, so this cost is allocated to the following key events.

- 6 months earlier than steel cutting: 20%
- Steel Cutting : 30%
- Keel Laying : 30%
- Launching : 10%
- Delivery : 10%
- Series Shipbuilding: Either give the entire cost to the first ship or distribute it equally among all ships.

Material Package Service Fee

Until the new shipyard develops its purchasing capabilities and domestic supply chain infrastructure, it is necessary to seek assistance from a leading global shipyard or engineering company for the procurement of shipbuilding materials and equipment. At this time, the following costs may be additionally included in the project's material costs.

- Transportation
- Service Engineering for Equipment Installation and Commissioning
- Packing for Shipping
- Procurement Services
- Insurance Fee
- Handling Charge

The cost distribution mentioned above applies using the same manner as in the case of the engineering package service fee.

Outsourcing Costs

This cost is a case of outsourcing production to a company outside the shipyard for a specific specialized product or mobilizing outsourced manpower in-house during periods of production overload. This is based on the premise that outsourcing costs are lower than conducting production activities using internal resources and/or facilities, while still maintaining price competitiveness.

Depreciation

Depreciation of tangible assets is allocated to expenses based on the useful life of each asset type mentioned earlier. It is typically distributed using the straight-line depreciation method after the shipyard's investment in the development stage is completed.

The economic effects of intangible assets, such as fees for new shipyard design and consulting, as well as technology transfer fees from a leading global shipyard, occur over a significant period of time, but for corporate valuation purposes, the useful life is set at 10 years.

Insurance Rates

In the case of a shipbuilding project, the shipbuilding insurance during the construction period amounts to approximately 0.5% of the ship price for a leading global shipyard, and the associated expenses are paid in a lump sum upon the ship's launch. In the case of an offshore project, the cost is approximately 0.75% of the ship price for a leading global shipyard, and the associated expenses are paid as a lump sum during the keel laying phase.

Other Manufacturing Costs

Among other manufacturing cost elements not covered above, variable costs in the form of direct costs are calculated as a percentage of material costs. Fixed costs in the form of indirect costs, are calculated based on a labor cost ratio.

In shipbuilding, the direct cost of other manufacturing expenses accounts for approximately 2±α% of the material cost, while the indirect cost of other manufacturing expenses amounts to approximately 20±α% of the labor cost.

2.3 Selling, General, and Administrative

Property Tax and Social Tax

Since each country has different rules for property tax on assets and social tax on wages, these taxes are subject to the tax laws of the respective countries.

Other Selling, General, and Administrative Expenses

Other SG&A expenses are calculated in the same way as other manufacturing costs. Direct expenses for other SG&A expenses typically account for approximately 0.5±α% of material costs, while indirect expenses make up approximately 9±α% of labor costs.

2.4 Earnings

Earnings Before Interest and Tax (EBIT)

EBIT, often expressed as Operating Income, is calculated as follows:

- EBIT = Gross Profit (= Revenue - Cost of Goods Sold) - Selling, General, and Administrative Expenses

Non-Operating Income

In the initial stage of a new shipyard, losses are generally inevitable. However, it is important to note that interest income, foreign currency exchange gains, and reversals of allowances for bad debts are not considered as operating income due to the short operating period. Furthermore, since the government provides various support policies such as tax benefits, loan interest rate reductions, and free land leases to attract and foster the shipbuilding industry, non-operating income is calculated based on the amount directly supported by the government.

Non-Operating Expenses

In the valuation stage of a new shipyard, non-operating expenses are not separately considered, except for interest on loans and fees related to technology transfer. Therefore, the interest rate of the respective country is applied to the entire loan for shipyard construction and facility investment, resulting in interest costs. Although the specific and clear calculation of the cost of technology transfer from a global leading shipyard is challenging, it would be more reasonable to consider it from the perspective of the opportunity cost associated with technology transfer from a global leading shipyard.

In general, various conflicts of interest may arise in mutual agreements on technology transfer fees, but as an example, the following can be assumed. However, it is important to consider the following factors and, above all, maintain a mutually beneficial and positive business attitude.

- Technical Transfer Fees = Σ [(Ship Price) x (EBIT Ratio of Global Shipyard) x (Annual Number of Ship Constructions at Local Shipyard) x (Intangible Asset Useful Life) x (Percentage of Shareholding of Global Shipyard)]

Ordinary Income

Ordinary income is calculated by deducting non-operating income and expenses from operating income, which serves as the foundation for corporate taxation.

2.5 Corporate Income Tax

The corporate income tax standards imposed on ordinary income also vary from country to country. In general, corporate income tax exemptions are granted for a specific period to attract foreign direct investment or promote the shipbuilding industry.

Some countries have laws that require ships transporting cargo within their own country to be built in domestic shipyards and/or provide government subsidies for shipbuilding at these shipyards. This is interpreted as a measure that fully takes into account the fact that if a new shipyard must achieve price competitiveness solely through productivity improvements, it will take a relatively long time to generate ordinary profits.

3 Valuation by Discounted Cash Flow

Balance Sheet

Balance sheets are typically prepared by accounting professionals, but depending on the payment terms for the price of the ship, there may be significant deviations from proper management of a shipyard's Operating Expenses (also known as OPEX). Therefore, it is essential to establish standardized commercial terms in new shipbuilding contracts, especially concerning payment terms. In the past, it was common to receive 20% of the ship price in advance and then receive the remaining balance in four installments of 20% for each key event. However, these days, payment terms tend to vary greatly depending on which party, the shipowner or shipbuilder, has the market initiative.

If the suitability of capital and loan size for facility investment is not considered, it may result in excessive interest costs being incurred as non-operating expenses. Therefore, it is necessary to evaluate the impact of capital size on the successful operation of a new shipyard.

Cash Flow

As is well known, the cash flow statement consists of cash flows from operating activities, cash flows from financing activities, and cash flows from investing activities. However, detailed accounting analysis techniques are not covered in this section.

Discounted Cash Flow Method

The DCF (Discounted Cash Flow) method for business valuation is a technique for analyzing the feasibility of a business. It involves calculating the Free Cash Flow (FCF), which is the surplus cash flow obtained by subtracting total investments from the future total cash flows. After calculating the FCF, it is then discounted to its present value using the appropriate discount rate, namely the Weighted Average Cost of Capital (WACC). This method is used to evaluate the feasibility of a business by converting its future cash flows into present value.

Here,

- FCF = Total Operating Cash Flow - Total Investments
  = After-Tax Operating Income + Depreciation - Total Investment
- Typical WACC Value: Approximately 12.4%
  (The actual calculation formula for WACC is very complex, but the average WACC value for the Korean shipbuilding industry is approximately 12.4%.)

4 Case Study

In order to ensure the long-term sustainability of a new shipyard, it is essential to conduct a comprehensive case study to analyze various conditions. The following are some of the key variables that need to be reviewed for the case study.

- Appropriate size and balance of capital and loans
- Sources of OPEX funds in the initial stages of shipyard operations (i.e., capital and loan coverage ratio)
- Investment funding plan for the shipyard development stage (i.e., capital, loan, government subsidy coverage ratio)
- Reduction of loan interest and amount of reduction
- Reduction or refund of value-added tax (VAT) incurred when investing in shipyard facilities
- Tax rates such as import duties, property taxes, income taxes, social taxes, value-added taxes, etc.
- Government or major shareholders providing support or subsidies for technology transfer fees

These conditions are set for each case, as shown in the example in Figure 51, and operating profit, non-operating expenses and taxes, and cumulative net profit are calculated for each case, as shown in Figure 52. As a result, it is necessary to analyze the point at which accumulated operating profit becomes positive. Furthermore, future free cash flow needs to be converted into present value discounted by WACC, as shown in Figure 53, to finally confirm the feasibility of the business in each case.

<table>
<tr><th colspan="3" rowspan="2">Categorization</th><th colspan="3">Option I</th><th colspan="3">Option II</th></tr>
<tr><th>Case I-1</th><th>Case I-2</th><th>Case I-3</th><th>Case II-1</th><th>Case II-2</th><th>Case II-3</th></tr>
<tr><td colspan="3">Option</td><td colspan="3">Technology Transfer Fee shall be distributed to each contract ;
- Technical document
- Training & OJT
- IT technology</td><td colspan="3">TTF shall be compensated by concept such as patent royalty ;
- License fee : xxx MUSD (In the time of TTF contract)
- TTF based on the local Shipyard's output :
Agreed rates on the time of vessel delivery at local shipyard</td></tr>
<tr><td rowspan="6">TAX</td><td colspan="2">Social</td><td colspan="3">0% (Normal tax rate after 10year operation)</td><td colspan="3">0% (Normal tax rate after 10year operation)</td></tr>
<tr><td colspan="2">Property</td><td colspan="3">0% (Normal tax rate after 10year operation)</td><td colspan="3">0% (Normal tax rate after 10year operation)</td></tr>
<tr><td colspan="2">Income</td><td colspan="3">Normal tax rate</td><td colspan="3">Normal tax rate</td></tr>
<tr><td colspan="2">Custom</td><td colspan="3">Imported goods of CAPEX : 10%</td><td colspan="3">Imported goods of CAPEX : 10%</td></tr>
<tr><td rowspan="2">VAT</td><td>Ship - building</td><td colspan="3">Refund</td><td colspan="3">Refund</td></tr>
<tr><td>CAPEX</td><td>No refund</td><td>Refund</td><td>- 1st : Govt. Subsidy
- 2nd : Refund</td><td>No refund</td><td>Refund</td><td>- 1st : Govt. Subsidy
- 2nd : Refund</td></tr>
<tr><td colspan="3">Interest & Subsidy</td><td colspan="3">- CAPEX & TTF : Refund 2/3 of posting interest rate after 3 months
- OPEX : Refund 2/3 of posting interest rate after 3 years</td><td colspan="3">- CAPEX & TTF : Refund 2/3 of posting interest rate after 3 months
- OPEX : Refund 2/3 of posting interest rate after 3 years</td></tr>
<tr><td rowspan="2">CAPEX</td><td colspan="2">1st Stage</td><td rowspan="2">- Loan : 80%
- Capital : 20%</td><td>- Loan : 80%
- Capital : 20%</td><td>- Govt. Subsidy: 80% (Repay after 10years)
- Capital : 20%</td><td rowspan="2">- Loan : 80%
- Capital : 20%</td><td>- Loan : 80%
- Capital : 20%</td><td>- Govt. Subsidy: 80% (Repay after 10years)
- Capital : 20%</td></tr>
<tr><td colspan="2">2nd Stage</td><td colspan="2">- Capital : 100%</td><td colspan="2">- Capital : 100%</td></tr>
<tr><td colspan="3">OPEX</td><td>- OPEX for 3 years : Loan</td><td colspan="2">- OPEX for 3 years : Capital</td><td>- OPEX for 3 years : Loan</td><td colspan="2">- OPEX for 3 years : Capital</td></tr>
<tr><td colspan="3">TTF Subsidy</td><td>None</td><td>- Subsidy by Major Shareholder (Repay after 10years)</td><td>None</td><td>None</td><td>- License fee : Subsidy by Major Shareholder (Repay after 10years)
- Remain TTF : Loan</td><td>None</td></tr>
</table>

Figure 51 Sample of Case Study Assumptions

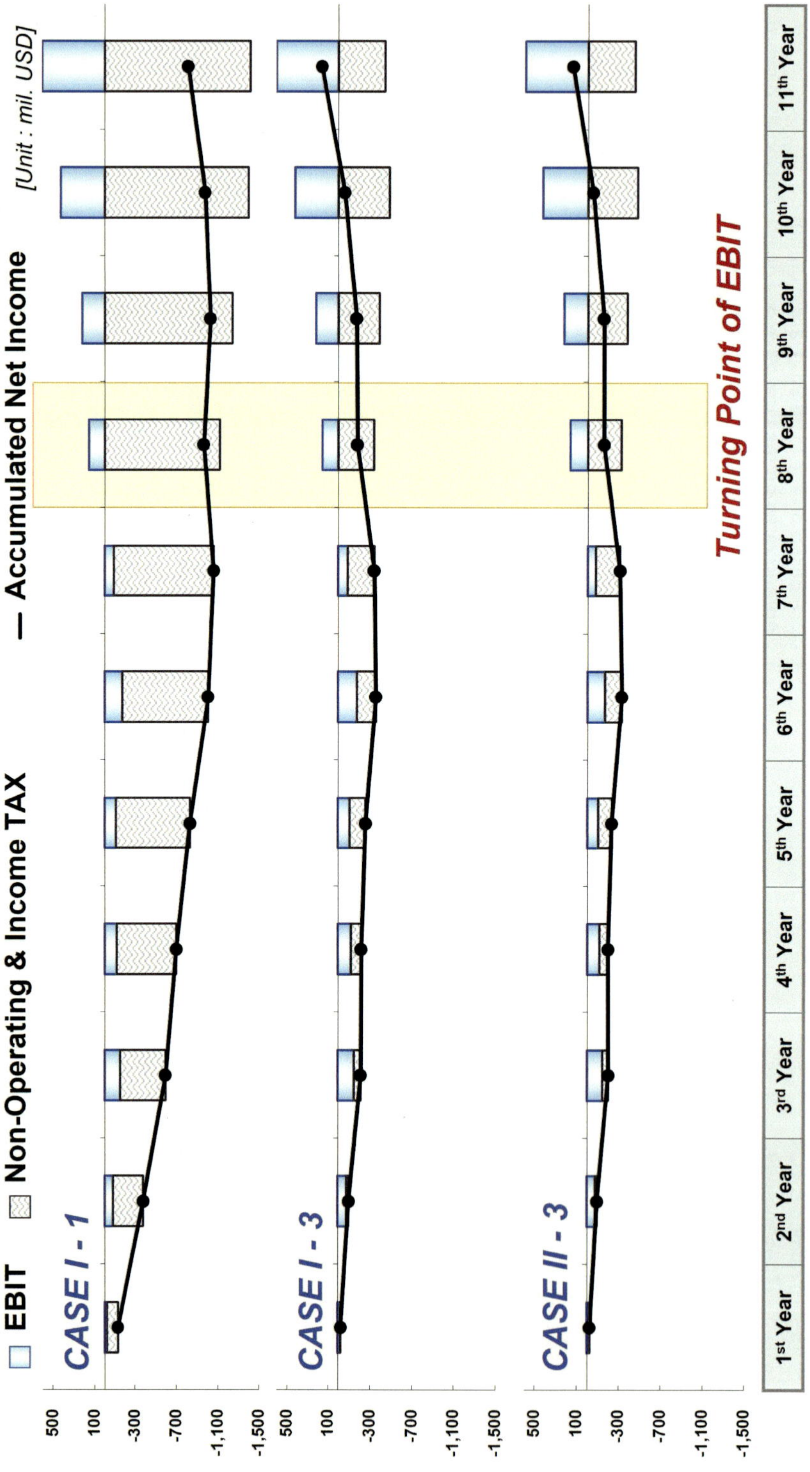

Figure 52 Sample of Accumulated Net Income Trend by Case Study Condition

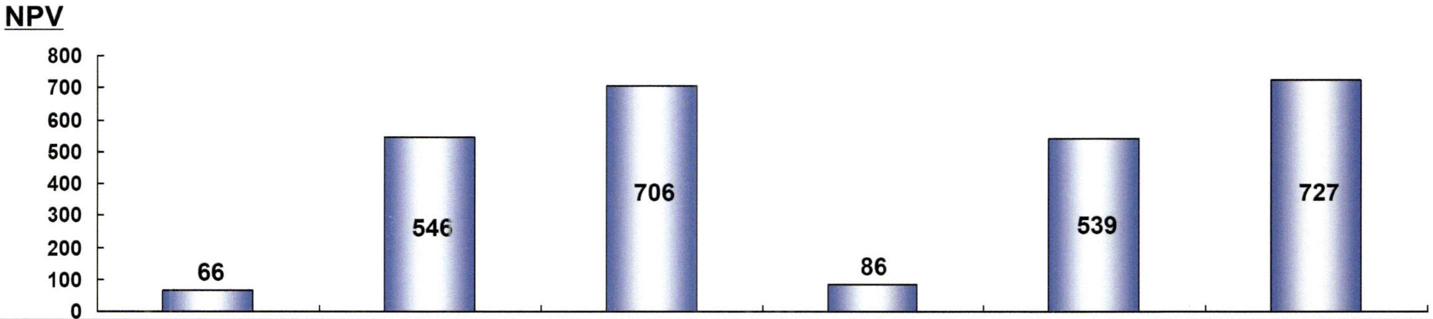

Case	Case I-1	Case I-2	Case I-3	Case II-1	Case II-2	Case II-3
IRR	13.1%	18.6%	20.6%	13.3%	18.8%	21.3%
Payback	10Year	9Year	9Year	10Year	9Year	9Year

** Remark : WACC 12.4% (Basis of Korean shipbuilding industry)*

Free Cash Flow

[Unit : mil. USD]

Year	1st Year	2nd Year	3rd Year	4th Year	5th Year	6th Year	7th Year	8th Year	9th Year	10th Year	After 11th Year
CASE I-1	-669	-735	-47	138	162	-131	148	310	176	692	2,509
CASE I-2	-598	-575	39	209	225	-84	195	347	223	729	2,546
CASE I-3	-563	-563	56	238	253	-55	219	370	249	752	2,569
CASE II-1	-640	-661	-44	120	147	-173	124	270	130	611	2,648
CASE II-2	-568	-505	36	186	204	-132	166	304	171	644	2,681
CASE II-3	-533	-489	59	220	238	-97	195	331	205	671	2,708

Figure 53 Sample Result of DCF Valuation

5 Further Considerations from the Valuation Result

As mentioned at the beginning of this book, due to the unique characteristics of the shipbuilding industry, it is nearly impossible for a new shipyard to generate operating profit and recover investment in a short period of time. Therefore, after evaluating all management conditions and business feasibility through the valuation process, it is necessary to explore the potential for coordination between policy authorities and various support policies for national key industries.

Ship Price and Materials Price in the Domestic Market

In order to expedite the normalization of management at new shipyards, it is necessary to provide a certain level of government subsidies for ship prices until comprehensive competitiveness is secured, or to support stable ship order activities through planned shipbuilding for national flag ships. In addition, if the domestic shipbuilding equipment supply chain is not competitive, this policy becomes even more necessary because the cost of equipment and material, which inevitably accounts for a significant portion of the ship's price, is not internationally competitive. This type of policy can be observed in various countries such as the United States, China, Brazil, and Angola, and Korea also implemented a planned shipbuilding policy in the beginning stages of the shipbuilding industry development.

Loan and Capital Scale

If there are no preferential loan interest rates or reductions in loan interest for the significant loan funds required for facility investment in a new shipyard, it is never easy to generate a positive net profit due to high non-operating costs. In this case, there is a need to reexamine the appropriate capital scale.

Taxation

In general, various tax benefits are typically provided to foster any industry until international competitiveness is achieved. Therefore, it is necessary to compare the major competitors in the global shipbuilding sector and analyze the various taxes associated with the industry. By providing tax benefits, even if only for a limited time, it will be crucial to enhance competitiveness. For such policy support, in some cases, the area where the shipyard is located is designated as a free trade industrial complex.

Government Subsidy

First, it is necessary to review various forms of government subsidies, such as preferential interest rates on facility investment loans, government funding for constructing industrial infrastructure outside of shipyards, and tax incentives for shipowners who order ships as part of ship price compensation. Furthermore, if a new shipyard effectively collaborates with government authorities to confirm the positive virtuous cycle effect of these government subsidies, extending beyond the shipbuilding industry to the national economy, the new shipyards will be able to stabilize their business in a short period of time, along with the justification for government subsidies.

Appendix A. Modern Shipbuilding General Processes (Video)

Accessing the link below will allow for viewing of the animation video, which facilitates understanding of *"Modern Shipbuilding Processes."*

Link:

Scan Me

Appendix B. Pre-Qualification Form (for Shipyard Construction)

I. Basic Information

1. Name of Company: ....
2. Registration Number: ....
3. Address: ....
4. Zip Code: ....
5. Telephone Number: ....
6. Fax Number: ....
7. Web Site: ....
8. President's Name: ....
9. Person to Contact (Full Name and Position):
   - Phone: ....
   - Email: ....
10. Branch Offices:

Location	*Telephone / Fax No.*
a.	....
b.	....
c.	....

11. Type of Business Entity

☐ Corporation ☐ Private Co. ☐ Consortium Partnership

12. Date of Incorporation:

13. Recent Sales Performance:

Year	Annual Sales (USD)	Gross Capital (USD)	Remarks

Please submit the following documents for the past three years.

a. Annual Report Including ESG Assessment

b. Financial Documents

c. Credit Rating Certificates

14. Project Performance Record

See and fill in Item No. III.

15. Safety Management

a. Do you have a written safety plan for field operations? ☐ Yes ☐ No

b. Lost Time Injury Frequency (LTIF)

= (Amount of Lost Time Injury / Annual Total Man-Hours
x 1,000,000 Man-Hours

Year	Lost Time Injury	Annual Total Man-Hour	LTIF	Remarks

16. Quality Management

a. ISO 9001:2015 Quality Management Systems ☐ Yes ☐ No

- Date of Certificate: ....

- Certification Authority: ....

b. ISO 14001:2015 Environmental Management Systems ☐ Yes ☐ No

- Date of Certificate: ....
- Certification Authority: ....

17. Project Management

a. ISO 21500:2012 Guidance on Project Management ☐ Yes ☐ No

b. Number of PMP Certificate: .... holders

18. Other Corporate Information Enclosures *(Please describe below)*:

a. ....

b. ....

c. ....

II. Construction Capability

1. Experienced Major Construction Activities (Please mark ✓)

Code	Categorization	Check	ID	Activities	Remarks
G	General	✓	0	General Contractor (All Work)	
				General Others	
C	Civil		0	Civil General	
			1	Earth Work	
			2	Piling Work	
			3	Civil Concrete Work	
			4	Soil Investigation	
			5	Surveying	
			6	Paving Work	
			7	Tunneling	
			8	Tensioning Work	
			9	Testing	
			10	Miscellaneous Steel Work	
			99	Others	
A	Architecture		0	Arch. General	
			1	Arch. Concrete Work	
			2	Arch. Structural Steel Work	
			3	Metal Work	
			4	Masonry Work	
			5	Plastering Work	
			6	Stonework	
			7	Tile Work	
			8	Waterproofing Work	
			9	Door & Window Work	

Code	Categorization	Check	ID	Activities	Remarks
A	Architecture		10	Finishing Work (Interior Work)	
			11	Ceiling Work	
			12	Curtain Wall Work	
			13	Roofing & Siding Work	
			14	Fence Work	
			15	Landscape Work	
			16	Fire Painting Work	
			17	Arch. Painting Work	
			18	Heating, Ventilation, and Air Conditioning	
			19	Sanitary System	
			20	Fire Protection System	
			21	Kitchen and Laundry	
			99	Others	
M	Mechanical		0	Mech. General	
			1	Mech. Structural Steel Work	
			2	Tank Erection Work	
			3	Piping Work	
			4	Insulation Work	
			5	Mech. Painting Work	
			6	Furnace & Boiler Erection Work	
			7	Refractory Lining Work	
			8	Heat Treatment Work	
			9	Non-Destructive Examination	
			10	Chemical Cleaning Work	
			11	Heavy Equipment Installation	
			12	Equipment Erection Work	

Code	Categorization	Check	ID	Activities	Remarks
M	Mechanical		13	Package Equipment	
			14	Firefighting Facility	
			99	Others	
E	Electrical Instrument		0	Elec. / Instrument General	
			1	Communication System	
			2	Electrical System	
			3	Elevator System	
			4	Fire Alarm System	
			5	Building Automation System	
			6	Instrument	
			99	Others	
T	Temporary		0	Temporary General	
			1	Temporary Architecture	
			2	Temporary Civil	
			3	Temporary Facilities	
			4	Temporary Electric	
			99	Others	
O	Others		1	Inland Transportation	
			2	Scaffolding	
			3	Security	
			4	Agent for Permit	
			5	Dismantling and Removal	
			6	Construction Equipment Supply	
			99	Others	

2. Construction Equipment Possessed and Operated Directly by the Company

Equipment	Manufacturer	Capacity	Prod. Year	Quantity	Remarks

3. Number of Permanent Employees

- **Total:** .... persons

  a. Management Staff: .... persons

  b. Administrators: .... persons

  c. Engineers: .... persons

  d. Draftsmen: .... persons

  e. Field Supervisors: .... persons

  f. Others: .... persons

4. Number of Skilled Laborers

Categorization	Permanent	Temporary	Total
1. Foreman (Supervisor)			
2. Pipe Welder			
3. Pipe Fitter			
4. Rigger			
5. Ironworker			
6. Mechanic			
7. Plate Welder			
8. Electrician			
9. Instrument Worker			
10. Insulation Worker			
11. Carpenter			
12. Painter			
13. Rebar Worker			
14. Bricker			
15. Plaster			
16. Concrete Worker			
17 . Equipment Operation Worker			
18. Truck Driver			
19. Semiskilled Laborer			
20. Common Laborer			
21. Others			
Total			

5. Number of Manuals, Standards, and Practices

Categorization	Manual	Standards	Practice	Total
1. Engineering & Design				
2. Engineering & Design Specification				
3. Manpower Control System				
4. Schedule Control System				
5. Progress Monitoring & Control System				
6. Cost Control System				
7. Safety Management				
8. Construction Procedure				
9. Quality Management				
10. Others				
Total				

6. Licenses and Certificates

Work and Service	License and Certificate ID Number	Contract Limitation (USD)	Remarks (Overseas License)

III. *Project Performance Record*

No.	Project Name (Location)	Project Period	Client	Contract Type	Project Scale	Contract Amount (USD)	Max. Manpower

** Note: Contract Type – **A**: Lump-Sum, **B**: Cost plus percentage fee, **C**: Cost plus fixed fee, **D**: Cost plus percentage with not-to-exceed limit, **E**: Unit price*

Epilogue

Finally, I am completing my writing, leaving behind the fear I had when I started, and most importantly, I am grateful to all of you who have been with me in the past.

First of all, I am deeply grateful to Daewoo Shipbuilding & Marine Engineering Co. Ltd. (currently known as "Hanwha Ocean"), which has been my home for the past 30 years since I first started my career. This company has allowed me to work with passion and dedication without wavering in my life. I consider myself fortunate to have started as a basic design engineer and gained experience in various aspects of shipyard operations. This has allowed me to develop a broad perspective as both a shipbuilding specialist and a generalist.

I would like to express my gratitude to Dong-Hyuk Park, my eternal guide and mentor. He has instilled in me a strong determination to pursue my ideals for the future with great ambition. We have spent a significant amount of time together throughout my professional career, which has greatly influenced my personal growth and development.

I will always remember my colleagues who worked overtime for a long time and engaged in extensive discussions to develop a logical approach for designing a new shipyard, despite the lack of previous reference materials. I would like to especially acknowledge the contributions of Hyo-Sang An, Jinsoo Park, Sujun Yoo, Chidon Kim, Seong-Mo Ku, Chan-Keun Kim, Jeong-Eun Lim, Untae Yeo, Sungwoo Bae, Sangyik Song, Je-Hyun Jung, and Ho-Jun Lee.

And in this process, I would like to express my gratitude to Hiroaki Tabata, Shigeru Daimon, Tsutomu Sudo, Hirofumi Kawase, and Taesuk Kim from Japan IHI. They have demonstrated honesty and exemplified the best global business spirit as technicians

through international collaboration. I would also like to thank Seo-Jin Han from GSI, Hyeng-Bo Shim and Donald Dongsup Kang from Korpec, Byeong-Uk Kim of Noory Architects, and Min-Chan Kim from DSNU in Korea for their unwavering support.

I would also like to thank Charlie Blakemore from BAE Systems in the UK, Antonio Dominguez from Navantia Shipyard in Spain, and Evgeny Kraynov, Igor Bukharev, and Kirill Zhukov from the Russian partner for demonstrating their passion as partners in the global shipbuilding cooperation project. Furthermore, I would like to extend special thanks to Taewan Kim and the other members of the Shipyard Development & Operation Consulting Team at DSEC Co. Ltd who collaborated with me on various consulting projects, such as productivity innovation at a shipyard in Spain and the development of a new shipyard in Russia.

Before the publication, I would like to express my gratitude to Professor Sung-Chul Shin from the Department of Naval Architecture at Pusan National University for his meticulous supervision of the manuscript. Lastly, I would like to express my gratitude to the individuals at DSEC Ltd. who supported the publication of this book, and to President Seo-Jin Han of GSI for their generous financial and moral support.

Based on my personal experience, I believe that there are several areas that lack content, including those that have not been covered yet and those that need to be approached from a different perspective. I sincerely hope that someone else will contribute to this in the future, and that the shipbuilding industry worldwide will continue to make progress. I also hope that this book will be helpful to countries that are endeavoring to establish a new shipbuilding industry.

Author Biography

Seungwoo Kang is a Republic of Korea national with extensive experience in shipyard core value chains, including quality assurance, design, production technology, production management, corporate planning, and procurement. With a strong ability to coordinate and integrate across a variety of fields, Kang has:

- 17 years of experience in basic engineering and project management for naval and passenger ships
- 3 years of experience in quality inspection and assurance
- 3 years of experience in production technology
- 4 years of experience in corporate planning and investment management
- 2 years of experience in material management for local supply chain procurement
- 5 years of experience in managing overseas affiliated companies and providing global shipyard consulting (for UK BAE Systems, Spain Navantia Shipyard, and Russia ZVEZDA)
- Served as a member of the Board of Directors for companies affiliated with DSME
- Served as a Senior Executive Vice President of DSEC, an engineering company affiliated with DSME.
- Served as the CEO of a heavy cargo transportation shipping company.

A Place
Where
Ships are Built

Guidelines for New Shipyard
Development and Design

- For inquiries regarding the purchase of the book,
  kindly contact the Pusan National University Press Center via email at **pnupress@pusan.ac.kr**

- Online purchasing options are available through the following platforms:
  - **Amazon US** (https://www.amazon.com)
  - **Aladdin US** (https://us.aladin.co.kr)
  - **Bandi & Lunis US** (https://www.bandibookus.com)